NAME #1350

ALSO BY ALAN LIGHTMAN

Origins

Ancient Light

Time for the Stars

Great Ideas in Physics

Einstein's Dreams

Good Benito

Dance for Two: Selected Essays

The Diagnosis

Reunion

A SENSE OF THE MYSTERIOUS

A SENSE OF THE
MYSTERIOUS

Science and the Human Spirit

ALAN LIGHTMAN

Pantheon Books, New York

All rights reserved under International and Pan-American Copyright
Conventions. Published in the United States by Pantheon Books,
a division of Random House, Inc., New York, and simultaneously
in Canada by Random House of Canada Limited, Toronto.

Pantheon Books and colophon are registered trademarks
of Random House, Inc.

Owing to limitations of space, all acknowledgments
for permission to reprint previously published material
may be found preceding the author's biography.

Library of Congress Cataloging-in-Publication Data

Lightman, Alan P., [date]
A sense of the mysterious : science and the human spirit /
Alan Lightman.
p.   cm.
ISBN 0-375-42320-6 (hardback)
1. Creative ability in science—United States.   2. Science—Philosophy.
3. Lightman, Alan P., [date]   4. Scientists—United States—20th
century—Biography.   I. Title.
Q172.5.C74L54 2005   501—dc22   2004052052

www.pantheonbooks.com

Book design by M. Kristen Bearse

Printed in the United States of America
First Edition
2 4 6 8 9 7 5 3 1

To my mentors in science,
William Gerace, Robert Naumann,
Martin Rees, and Kip Thorne

CONTENTS

A SENSE OF THE MYSTERIOUS

A SENSE OF THE MYSTERIOUS

EVER SINCE I WAS a young boy, my passions have been divided between science and art. I was fortunate to make a life in both, as a physicist and a novelist, and even to find creative sympathies between the two, but I have had to live with a constant tension in myself and a continual rumbling in my gut.

In childhood, I wrote dozens of poems. I expressed in verse my questions about death, my loneliness, my admiration for a plum-colored sky, my unrequited love for fourteen-year-old girls. Overdue books of poetry and stories littered my second-floor bedroom. Reading, listening, even thinking, I was mesmerized by the sounds and the movement of words. Words could be sudden, like *jolt*, or slow, like *meandering*. Words could be sharp or smooth, cool, silvery, prickly to touch, blaring like a trumpet call, fluid, pitter-pattered in rhythm. And, by magic, words could create scenes and emotions. When my grandfather died, I buried my grief in writing a poem, which I showed to my grandmother a month later. She cradled my face with her veined hands and said, "It's beautiful," and then began weeping all over again. How could marks on a white sheet of paper contain such power and force?

Between poems, I did scientific experiments. These I conducted in the cramped little laboratory I had built out of a storage closet in my house. In my homemade alchemist's den, I hoarded resistors and capacitors, coils of wire of various thicknesses and grades, batteries, switches, photoelectric cells, magnets, dangerous chemicals that I had secretly ordered from unsuspecting supply stores, test tubes and Petri dishes, lovely glass flasks, Bunsen burners, scales. I delighted in my equipment. I loved to build things. Around the age of thirteen, I built a remote-control device that could activate the lights in various rooms of the house, amazing my three younger brothers. With a thermostat, a lightbulb, and a padded cardboard box, I contructed an incubator for the cell cultures in my biology experiments. After seeing the movie *Frankenstein,* I built a spark-generating induction coil, requiring tedious weeks upon weeks of winding a mile's length of wire around an iron core.

In some of my scientific investigations, I had a partner, John, my best high-school friend. John was a year older than I and as skinny as a strand of 30 gauge wire. When he thought something ironic, he would let out a high-pitched, shrill laugh that sounded like a hyena's. John did not share my interest in poetry or the higher arts. For him, all that was a sissyish waste of calories. John was all practicality. He wanted to seize life by the throat and pull out the answer. As it turned out, he was a genius with his hands. Patching together odds and ends from his house, he could build anything from scratch. John never saved the directions that came with new parts, he never drew up detailed schematic diagrams, and his wiring wandered drunkenly around the circuit board, but he had the magic touch, and when he would sit down cross-legged on the floor of his room and begin fiddling, the transistors hummed. His inventions were not pretty, but they worked, often better than mine.

Weekends, John and I would lie around in his room or mine, bored, listening to Bob Dylan records, occasionally thinking of things to excite our imaginations. Most of our friends filled their weekends with the company of girls, who produced plenty of excitement, but John and I were socially inept. So we listened to Dylan and read back issues of *Popular Science*. Lazily, we perused diagrams for building wrought-iron furniture with rivets instead of welded joints, circuits for fluorescent lamps and voice-activated tape recorders, and one-man flying machines made from plastic bleach bottles. And we undertook our ritual expeditions to Clark and Fay's on Poplar Avenue, the best-stocked supply store in Memphis. There, we squandered whole Saturdays happily adrift in the aisles of copper wire, socket wrenches, diodes, oddly shaped metallic brackets that we had no immediate use for but purchased anyway. Clark and Fay's was our home away from home. No, more like our temple. At Clark and Fay's, we spoke to each other in whispers.

Our most successful collaboration was a light-borne communication device. The heart of the thing was a mouthpiece made out of the lid of a shoe polish can, with the flat section of a balloon stretched tightly across it. Onto this rubber membrane we attached a tiny piece of silvered glass, which acted as a mirror. A light beam was focused onto the tiny mirror and reflected from it. When a person talked into the mouthpiece, the rubber vibrated. In turn, the tiny mirror quivered, and those quiverings produced a shimmering in the reflected beam, like the shimmering of sunlight reflected from a trembling sea. Thus, the information in the speaker's voice was precisely encoded onto light, each rise and dip of uttered sound translating itself into a brightening or dimming of light. After its reflection, the fluttering beam of light traveled across John's messy bedroom to our receiver, which was built from largely off-the-shelf stuff: a photocell to convert varying intensities of light into varying intensities of electrical current, an amplifier, and a microphone to convert electrical current into sound. Finally, the original voice was reproduced at the other end. Like any project that John was involved in, our communication device looked like a snarl of spare parts from a junkyard, but the thing worked.

It was with my rocket project that my scientific and artistic proclivities first collided. Ever since the launch of *Sputnik* in October 1957, around my ninth birthday, I had been entranced with the idea of sending a spacecraft aloft. I imagined the blast-off, the uncoiling plume of smoke, the silvery body of the rocket lit by the sun, the huge acceleration, the beautiful arc of the trajectory in the sky. By the age of fourteen, I was experimenting with my own rocket fuels. A fuel that burned too fast would explode like a bomb; a fuel that burned too slow would smolder like a barbecue grill. What seemed to work best was a mixture of powdered charcoal and zinc, sulfur, and potassium nitrate. For the ignition, I used a flashbulb from a Brownie camera, embedded within the fuel chamber. The sudden heat from the bulb would easily start the combustion, and the bulb could be triggered by thin wires trailing from the tail of the rocket to the battery in my control center, a hundred feet away. The body of the rocket I built from an aluminum tube. The craft had red tail fins. It was beautiful. For a launching pad, I used a V-shaped steel girder, pointed skyward at the appropriate angle and anchored in a wooden Coca-Cola crate filled with concrete.

I invited my awed younger brothers and several friends

from the neighborhood to attend the rocket launch, which took place one Sunday at dawn at Ridgeway Golf Course. John, who was not in the slightest a romantic and didn't see anything useful about rockets, elected to stay in his bed and sleep. But even so, I had a good audience. Because I had estimated from thrust and weight calculations that my rocket might ascend a half mile into space, some of the boys brought binoculars. From my control center, I called out the countdown. I closed the switch. Ignition. With a flash and a whoosh, the rocket shot from its pad. But after rising only a few hundred feet, it did a sickening swerve, spun out of control, and crashed. The fins had come off. With sudden clarity, I remembered that instead of riveting the fins to the rocket body, as I should have, I had glued them on. To my eye, the rivets had been far too ugly. How I thought that mere glue would hold under the heat and aerodynamic force, I don't know. Evidently, I had sacrificed reality for aesthetics. John would have been horrified.

Later, I learned that I was not the first scientist for whom beauty had ultimately succumbed to reality. Aristotle famously proposed that as the heavens revolved about the earth, the planets moved in circles. Circles because the circle is the simplest and most perfect shape. Even when astronomers discovered that the planets sometimes zigzagged in their paths, showing that they couldn't remain in simple orbits, scientists remained so enthralled with the circle that they decided the planets must move in little circles attached to big circles. The circle idea was lovely and appealing. But it was proved wrong by the careful observations of Brahe and Kepler in the late sixteenth and early seventeenth centuries. Planets orbit in ellipses, not circles. Equally beautiful was the idea, dating from the 1930s, that all phenomena of nature should be completely identical if right hand and left hand were reversed, as if reflected in a mirror. This elegant idea, called "parity conservation," was proven wrong in the late 1950s by the experiments proposed by Lee and Yang, showing that some subatomic particles and reactions do not have identical mirror-image twins. Contrary to all expectations, right- and left-handedness are not equal.

When my scientific projects went awry, I could always find certain fulfillment in mathematics. I loved mathematics just as I loved science and poetry. When my math teachers assigned homework, most other students groaned and complained, but I relished the job. I would save my math problems for last, right before bedtime, like bites of chocolate cake awaiting me after a long and dutiful meal of history and Latin. Then I would devour my cake. In geometry, I loved drawing the diagrams; I loved finding the inexorable and irrefutable relations between lines, angles, and curves. In algebra, I loved the idea of abstraction, letting x's and y's stand for the number of nickels in a jar or the height of a building in the distance. And then solving a set of connected equations, one logical step after another. I loved the shining purity of mathematics, the logic, the precision. I loved the certainty. With mathematics, you were guaranteed an answer, as clean and crisp as a new twenty-dollar bill. And when you had found that answer, you were right, unquestionably right. The area of a circle is πr^2. Period.

Mathematics contrasted strongly with the ambiguities and contradictions in people. The world of people had no certainty or logic. People confused me. My mother sometimes said cruel things to me and my brothers, even though I felt that she loved us. My aunt Jean continued to drive recklessly and at great speeds, even though everyone told her that she would kill herself in an automobile. My uncle Edwin asked me to do a mathematical calculation that would help him run the family business with more efficiency, but when I showed him the result he brushed it aside with disdain. Blanche, the dear woman who worked forever for our family, deserted her husband after he abused her and then talked about him with affection for years. How does one make sense out of such actions and words?

A long time later, after I became a novelist, I realized that the ambiguities and complexities of the human mind are what give fiction and perhaps all art its power. A good novel gets under our skin, provokes us and haunts us long after the first reading, because we never fully understand the characters. We sweep through the narrative over and over again, searching for meaning. Good characters must retain a certain mystery and unfathomable depth, even for the author. Once we see to the bottom of their hearts, the novel is dead for us.

Eventually, I learned to appreciate both certainty and uncertainty. Both are necessary in the world. Both are part of being human.

IN COLLEGE, I made two important decisions about my career.

First, I would put my writing on the back burner until I became well established in science. I knew of a few scientists who later became writers, like C. P. Snow and Rachel Carson, but no writers who later in life became scientists. For some reason, science—at least the creative, research side of science—is a young person's game. In my own field, physics, I found that the average age at which Nobel Prize winners did their prize-winning work was only thirty-six. Perhaps it has something to do with the focus on and isolation of the subject. A handiness for visualizing in six dimensions or for abstracting the motion of a pendulum favors an agility of mind but apparently requires little knowledge of the human world. By contrast, the arts and humanities require experience with life and the awkward contradictions of people, experience that accumulates and deepens with age.

Second, I realized that I was better suited to be a theorist than an experimentalist. Although I loved to build things, I simply did not have the hands-on dexterity and practical talents of the best students. My junior-year electronics project caught fire when I plugged it in. My senior thesis project, a gorgeous apparatus of brass fittings and mylar windows designed to measure the half-life of certain radioactive atoms, was sidelined on the lab bench instead of being installed in the cyclotron for a real experiment. I never did believe the thing would actually work. And apparently neither did my professor, Robert Naumann, who kindly gave me high marks for my endless drawings of top views and side views and calculations of solid angles and efficiencies. In various ways, I was making the self-discovery that I was destined to be a theorist, a scientist who worked with abstractions about the physical world, ideas, mathematics. My equipment would be paper and pencil.

It was in college, also, that I realized there were other young people as excited about science as I was. In my sophomore year, I came under the wing of a renegade assistant professor of physics named Bill Gerace. Gerace had had some vague disagreements with the powers that be and had responded by forming his own tiny university, a group of a half-dozen students who lectured one another on physics and worked on independent projects under Professor Gerace's supervision and constant encouragement. We were all given desks in Gerace's "office," a room in the dark basement of the physics building. For us chosen ones, that room was a palace. Many of the beauties of thermodynamics, wave functions, and other magical parts of physics first unfolded for me in that basement room. I will never forget how I was recruited to join this university within a university. Gerace, a lab instructor during my first physics course, walked up to me one day at the end of an experiment and furtively assigned me a theoretical problem not connected with the lab: if a frictionless bug sits on the outer rim of a frictionless clock, starting at the twelve o'clock position, and begins sliding clockwise, at what angle (or hour mark) will the bug fall off? When I brought in the answer the next day, Gerace got a big smile on his face and told me that I was a physicist.

A year or two after college, I had my first true experience with original research. It was an experience that I can compare only to my first love affair. At the time, I was twenty-two years old, a graduate student in physics at the California Institute of Technology. My thesis adviser at Caltech was Kip Thorne, only thirty himself but already a full professor. Kip had grown up in Mormon Utah but had completely acclimatized to the hip zone of California in the early 1970s. He sported long red hair, starting to thin, a red beard, sandals, loose kaftanlike shirts splotched with colors, sometimes a gold chain around his neck. Freckled, lean limbed, wiry. And brilliant. His specialty was the study of general relativity, Einstein's theory of gravity. In fact, there was at this time a renaissance of interest in Einstein's arcane theory because astronomers had recently discovered new objects in space, such as neutron stars, that had enormous gravity and would require general relativity for a proper understanding.

One of Kip's programs was to compare general relativity to other modern theories of gravity. And it was in that program that he assigned me my first research problem. I was supposed to show, by mathematical calculation, whether a particular experimental result required that gravity be geometrical. The known experimental result was that all objects fall under gravity with the same acceleration. Drop a book and a cannonball from the same height, and they will hit the floor at the same time, if air resistance is small. That result says something profound about the nature of gravity. By "geometrical" Kip meant that gravity could be described completely as a warping of space. In such a picture, a mass like the sun acts as if it were a heavy weight sitting on a stretched rubber sheet; orbiting planets follow along the sagging surface of the sheet. In the early 1970s, some modern theories of gravity, such as Einstein's general relativity, were geometrical. Some were not. To be "geometrical," to be equivalent to a bending of space, a theory had to have a particular mathematical form. So my project amounted to writing down on a piece of paper the equations representing a giant umbrella theory of gravity, a "theory of theories" that encompassed many different possible theories, next imposing the restriction that all objects fall with the same acceleration, and then finding out whether that restriction was sufficiently powerful to rule out all nongeometrical theories.

I was both thrilled and terrified by my assignment. Until this point in my academic life, my theoretical adventures had consisted mainly of solving homework problems. With homework problems, the answer was known. If you couldn't solve the problem yourself, you could look up the answer in the back of the book or ask a smarter student for help. But this research problem with gravity was different. The answer wasn't known. And even though I understood that my problem was inconsequential in the grand sweep of science, it was still original research. No one would know the answer until I found it. Or failed to find it.

After an initial period of study and work, I succeeded in writing down all the equations I thought relevant.

Then I hit a wall. I knew something was amiss, because a simple result at an early stage in the calculation was not coming out right. But I could not find my error. And I didn't even know what kind of error. Perhaps one of the equations was wrong. Or maybe the equations were right but I was making a silly arithmetic mistake. Or perhaps the conjecture was false but would require an especially devious counterexample to disprove it. Day after day, I checked each equation, pacing back and forth in my little windowless office, but I didn't know what I was doing wrong. This confusion and failure went on for months. For months, I ate, drank, and slept my research problem. I began keeping cans of tuna fish in a lower drawer of the desk and eating meals in my office.

Then one morning, I remember that it was a Sunday morning, I woke up about five a.m. and couldn't sleep. I felt terribly excited. Something strange was happening in my mind. I was thinking about my research problem, and I was seeing deeply into it. I was seeing it in ways I never had before. The physical sensation was that my head was lifting off my shoulders. I felt weightless. And I had absolutely no sense of my self. It was an experience completely without ego, without any thought about consequences or approval or fame. Furthermore, I had no sense of my body. I didn't know who I was or where I was. I was simply spirit, in a state of pure exhilaration.

The best analogy I've been able to find for that intense feeling of the creative moment is sailing a round-bottomed boat in strong wind. Normally, the hull stays down in the water, with the frictional drag greatly limiting the speed of the boat. But in high wind, every once in a while the hull lifts out of the water, and the drag goes instantly to near zero. It feels like a great hand has suddenly grabbed hold and flung you across the surface like a skimming stone. It's called planing.

So I woke up at five to find myself planing. Although I had no sense of my ego, I did have a feeling of rightness. I had a strong sensation of seeing deeply into the problem and understanding it and knowing that I was right—a certain kind of inevitability. With these sensations surging through me, I tiptoed out of my bedroom, almost reverently, afraid to disturb whatever strange magic was going on in my head, and I went to the kitchen. There, I sat down at my ramshackle kitchen table. I got out the pages of my calculations, by now curling and stained. A tiny bit of daylight was starting to seep through the window. Although I was oblivious to myself, my body, and everything around me, the fact is I was completely alone. I don't think any other person in the world would have been able to help me at that moment. And I didn't want any help. I had all of these sensations and revelations going on in my head, and being alone with all that was an essential part of it.

Somehow, I had reconceptualized the project, spotting my error of thinking, and begun anew. I'm not sure how this rethinking happened, but it wasn't by going from one equation to the next. After a while at the kitchen table, I solved my research problem. I had proved that the conjecture was true. The equal acceleration of the book and the cannonball does indeed require that gravity be geometrical. I strode out of the kitchen, feeling stunned and powerful. Suddenly I heard a noise and looked up at the clock on the wall and saw that it was two o'clock in the afternoon.

I was to experience this creative moment again, with other scientific projects. But this was my first time. As a novelist, I've experienced the same sensation. When I suddenly understand a character I've been struggling with, or find a lovely way of describing a scene, I am lifted out of the water, and I plane. I've read the accounts of other writers, musicians, and actors, and I think that the sensation and process are almost identical in all creative activities. The pattern seems universal: The study and hard work. The prepared mind. The being stuck. The sudden shift. The letting go of control. The letting go of self.

I learned many things about science from Kip. One of the most important was the concept of the "well-posed problem." A well-posed problem is a problem that can be stated with enough clarity and definiteness that it is guaranteed a solution. Such a solution might require ten years, or a hundred, but there should be a definite solution. While it is true that science is constantly revising itself to respond to new information and ideas, at any moment scientists are working on well-posed problems.

I often think of Kip's idea of the well-posed problem as closely related to Karl Popper's notion of what makes a scientific proposition. According to Popper, who was an important early-twentieth-century British philosopher of science, a scientific proposition is a statement that can in principle be proven false. Unlike with mathematics, which exists completely within its own world of logical abstraction, you can never prove a scientific proposition or theory true because you can never be sure that tomorrow you might not find a counterexample in nature. Scientific theories are just simplified models of nature. Such a model might be mathematically correct, but its beginning premises might not be in sufficient accord with physical reality. But you can certainly prove any scientific theory false. You can find a counterexample, an experiment that disagrees with the theory. And, according to Popper, unless you can at least imagine an experiment that might falsify a theory, that theory or statement is not scientific.

In direct and indirect ways, Kip emphasized to his students that we should not waste time on problems that weren't well-posed problems. I have since come to understand that there are many interesting problems that are not well posed in the Popper or Thorne sense. For example: Does God exist? Or, What is love? Or, Would we be happier if we lived a thousand years? These questions are terribly interesting, but they lie outside the domain of science. Never will a physics student receive his or her degree working on such a question. One cannot falsify the statement that God exists (or doesn't exist). One cannot falsify the statement that we would be happier (or not happier) if we lived longer. Yet these are still fascinating questions, questions that provoke us and bring forth all kinds of creative thought and invention. For many artists and humanists, the question is more important than the answer. One of my favorite passages from Rilke's *Letters to a Young Poet* is this: "We should try to love the questions themselves, like locked rooms and like books that are written in a very foreign tongue." Science is powerful, but it has limitations. Just as the world needs both certainty and uncertainty, the world needs questions with answers and questions without answers.

Another thing I learned from Kip, more a matter of personal style, was generosity. Kip bent over backward to give credit first to his students. He would put his name last on joint papers, he would heap praise on his students at public lectures. Kip was well aware of his strengths, but he was modest at the same time, and he was deeply generous in his heart. I believe that he inherited these virtues from his own thesis adviser at Princeton, John Wheeler. Wheeler, in turn, absorbed much of his personal style from his mentor, the great atomic physicist Niels Bohr, in Copenhagen. In a sense, I was a "great-grand-student" of Bohr's.

Three Caltech professors served on my thesis committee, charged with examining me at my final thesis defense. Richard Feynman was one of the three. For some years, Feynman had taken an interest in Kip's students and, every couple of months, would go to lunch with us and pepper us with questions about the latest findings in gravitational waves or black holes or some other topic in general relativity. At my thesis defense, I stood at a blackboard in a small room while these guys sat comfortably and asked me questions. Feynman asked the first two questions. His first question was rather easy, and I answered it without too much trouble. His second question was just a little beyond my reach. I struggled with it; I went sideways and backwards; I circled around. Finally, after about twenty minutes of fumbling at the blackboard, I managed an answer. Feynman asked no more questions. Later, I realized that with his two questions he had precisely bracketed my ability. He had launched two artillery shells at me, one falling short, one long, and he knew exactly where I was in the intellectual landscape of physics.

I VIVIDLY REMEMBER a scene from sometime in 1975. It takes place during my two years as a postdoctoral fellow at Cornell. I am sitting on a couch in Edwin Salpeter's house. Ed, suffering from one of his recurring back problems, lies on the floor. From that low vantage, he is helping me think through a problem involving stars being ripped apart and consumed by a giant black hole. It is a theoretical problem, of course.

At this time, Ed would have been about fifty years old. He was widely regarded as one of the two or three greatest theoretical astrophysicists in the world. His most famous work, done in the 1950s, involved the theoretical recipe for how helium atoms in stars can combine to make carbon, and then heavier elements beyond that. It is believed that all of the chemical elements in the universe heavier than the two lightest, hydrogen and helium, were forged at the centers of stars. Ed and his colleagues showed how that process was possible. Among some of his other accomplishments, Ed calculated how many stars should be created in each range of mass—a sort of birth-weight chart for newborn stars.

When I first arrived at Cornell, in the fall of 1974, Ed immediately dragged me out to the tennis courts to find out what I was made of. I was a fair tennis player. After a number of exhausting matches over the season, we were approximately tied, but Ed could not refrain from quietly gloating whenever he beat me. And I could see that same gentlemanly but competitive edge in his science. He didn't like to lose.

On and off the tennis court, Ed dressed in tattered short-sleeve sports shirts. These, combined with his loafers and stylishly long hair and faint accent from his Austrian roots, gave him an air of casual elegance. But Ed was enormously serious about his physics. When he was talking about a physics problem, he would sometimes stop, turn his head, and just stare off into space for a few moments, and you knew that he was delving into deeper layers of thought.

What I found most brilliant about Ed was his physical intuition. He could visualize a physical problem and almost feel his way to the core of it, all in his head. This ability arose from his vast knowledge of physics and astronomy and his talent for making analogies from one subject to another. Many of the greatest scientists have had this talent for analogies. Planck compared the inside surface of a container to a collection of springs with different oscillation frequencies. Bohr compared the nucleus of an atom to a drop of liquid.

So we're in Ed's living room, me on the couch, Ed on his back on the floor, some kind of classical music floating in from the next room, and Ed draws an analogy between stars being swallowed by the big black hole and a drunk wandering on a street that has an uncovered sewer hole. If a star comes too close to the black hole it will be destroyed, just as if the drunk stumbles to the sewer hole he will fall in. Each star, in each orbit around the central black hole, is given a random jostle by the gravity of the other stars, just as the drunk takes a random step every second. Such random steps can lead a star, or a drunk, to fall into the hole. The stars bump about in two-dimensional "angular-momentum space," just as the drunk wanders around on a two-dimensional street. The critical question, Ed announces from the floor, is whether each random step of the drunk is bigger or smaller than the diameter of the hole. With this insight, I and the other postdoctoral fellow collaborating with me on the problem can now work out the details. The result will be a prediction for the Hubble Space Telescope, more than a decade away. Ed asks if I would please bring him a cup of tea. He has other things to think about this morning.

Some months later, I had a severe emotional upheaval with a different scientific project. I was working on the arrangement of stars in a globular cluster. A globular cluster is a congregation of about a hundred thousand stars, all orbiting one another under their mutual gravitational attraction. There are about a hundred globular clusters in our galaxy. Through a telescope, a globular cluster appears as a beautiful, shining ball of light. Imagine: a hundred thousand stars all concentrated together in a tight ball, whizzing about like angry bees in a bee's nest.

Since about 1970, astrophysicists had begun to simulate the structure and evolution of globular clusters on a computer. You feed the computer the initial position of a lot of points, each representing a star or group of stars, you put in the effects of gravity, each point gravitationally attracting all the others, and you let the computer tell you what happens in time. In a sense, the computer is doing an experiment for you. Each minute of computer time might represent a million years for the globular cluster. One of the findings of these "experiments" was that the simulated globular clusters begin collapsing. The inner stars lose energy and move closer to the center, while the outer stars gain energy and move farther from the center. For extra gratification, there were even observations of actual globular clusters in space, observations suggesting that some globular clusters may indeed have undergone such a collapse.

Many of the computer simulations had been done with the simplification that all stars have the same mass. I wanted to investigate what happens under the more realistic assumption that there is a range of masses of stars. But instead of doing a computer simulation, which is extremely time-consuming to set up and costly to run,

I found an approximate way to attack the problem using only pencil and paper. As I suspected, having a range of masses of stars made the cluster collapse even sooner and faster.

While in the final stages of writing up my results for publication, I strolled into the astronomy library to complete my list of references to previous work. And there, to my horror, I discovered a brand-new issue of *Astrophysics and Space Science* in which two Japanese scientists had solved the same problem. With my pulse racing, I checked their results against mine. Our figures and graphs agreed to within three decimal places. I had been scooped! Of course, most people get scooped at various times in their lives if they're working on anything at all interesting. But this was the first time for me.

I experienced a complex set of reactions. I was embarrassed. I was humiliated. I grieved the loss of several months of my time. I worried whether the wasted effort would compromise my chances for an assistant professorship. But then, another emotion began working its way through my body. Amazement. I was utterly amazed that people on the other side of the planet, with no correspondence between us, no comparing of notes, had decided to solve the same problem and had gotten the same answer to three decimal places. There was something wonderful and thrilling about that. Here was powerful evidence of a thing—part science, part mathematics—that exists outside of our own heads. Presumably, Martians also would have gotten the same answer to three decimal places had they chosen to work on the same problem. There was terrible precision in the world.

After this feeling of awe at the terrible precision and exactness of the world, I began to experience another emotion: irrelevancy. If the physical universe was reducible to precise equations with precise answers to three decimal places (and more), then why was I, as a particular and unique person, needed to find those answers? For the globular cluster problem with multiple masses, Saito and Yoshizawa had found the answer before me. If neither they nor I had found the answer, then in another month or another year somebody else would have found it. Science is not an occupation for a person who wants to make a mark as an individual, accomplishing something only that individual can do. In science, it is the final measured number or the final equation that matters most. The important experiments are those that can be repeated over and over again, in laboratories all over the world, with the same results. The important equations are those that can be rederived by anyone with sufficient training. If Heisenberg and Schrödinger hadn't formulated quantum mechanics, then someone else would have. If Einstein hadn't formulated relativity, then someone else would.

If Watson and Crick hadn't discovered the double-helical structure of DNA, then someone else would. Science brims with colorful personalities, but the most important thing about a scientific result is not the scientist who found it but the result itself. Because that result is universal. In a sense, that result already exists. It is only *found* by the scientist. For me, this impersonal, disembodied character of science is both its great strength and its great weakness.

I couldn't help comparing the situation to my other passion, the arts. In the arts, the individual is the essence. Individual expression is everything. You can separate Einstein from the equations of relativity, but you cannot separate Beethoven from the *Moonlight Sonata*. No one will ever write *The Tempest* except Shakespeare or *The Trial* except Kafka.

I loved the grandeur, the power, the beauty, the logic and precision of science, but I also ached to express something of myself—my individuality, the particular way that I saw the world, my unique way of being. On that day in the Cornell library, as I feverishly turned the pages of *Astrophysics and Space Science,* I learned something about science, and I also learned something about myself. I would continue following my passion in science, but I could no longer suppress my passion for writing.

FINALLY, IN THE EARLY 1980S, I began writing essays. For some years I had been publishing poems in small literary magazines. The essay gave me the greater flexibility I wanted. With an essay, I could be informative, poetic, philosophical, personal. And, at a time when most of my self-identity and confidence were still based on my achievements as a scientist, with the essay I could connect my scientific and artistic interests. I would come home in the evening, elated from a day of research at the Harvard-Smithsonian Center for Astrophysics, and ponder an essay.

One of my first essays concerned Joseph Weber, a distinguished professor of physics at the University of Maryland. Weber had pioneered the first gravitational wave detectors. He had also become somewhat of an outcast in the scientific community because he claimed to have seen gravitational waves when no one else could. When you shake an electrical charge, it emits waves of electricity and magnetism that travel through space at the speed of light. Likewise, Einstein's general relativity predicted that when you shake a mass of any kind, whether electrically charged or not, it emits gravitational waves, waves of oscillating gravity that travel through space also with the speed of light. Hypothetically, the strongest sources of such waves would be cataclysmic cosmic events, such as the collision of black holes in space.

How does one observe a gravitational wave? When a gravitational wave strikes a mass, it causes that mass to expand and contract like a working bellows pump. Gravitational waves, however, are fantastically weaker than electromagnetic waves. A typical expansion or contraction expected for a cosmic gravitational wave might be one part in 10^{21} or smaller, corresponding to a thousand-mile-long ruler changing its length by the width of a single atomic nucleus. Consequently, while a high-school student can build a crystal radio set to detect electromagnetic waves, gravitational waves require extraordinarily sensitive equipment to measure them.

In 1960, when no one else was dreaming of detecting gravitational waves, Weber conceived of the idea of a resonant cylinder, a metallic cylinder that would ring like a bell (but an extremely soft bell) when struck by a gravitational wave. One of the problems in building such a resonant cylinder, or any detector, is that it is always expanding and contracting a little bit from tiny random disturbances, such as a truck turning a corner a half mile away. It is extremely difficult to discriminate such noise from the minuscule motions expected from a gravitational wave. So you build two cylinders, thousands of miles apart, and monitor them closely. If both of them begin softly ringing in precisely the same way at the same time, then perhaps they've just been struck by a gravitational wave.

In the early 1960s, Weber built such cylinders, the first one located at the University of Maryland, near Washington, D.C., the second at Argonne National Laboratory, near Chicago. Each cylinder had a length of five feet, a diameter of about two feet, and a weight of about three thousand pounds. In 1968, Weber began reporting the observation of simultaneous oscillations of his two cylinders. He claimed to have discovered the first gravitational waves.

In the following decade, other groups of scientists attempted to duplicate Weber's results. They built their own cylinders, hooked them up to their own piezo-electrical crystals to measure minute oscillations, compared their own charts of the oscillations in time. No one saw oscillations of the magnitude claimed by Weber, and no one saw simultaneous oscillations of their cylinders except what would be expected by chance. In fact, other detectors were built with a hundred times more sensitivity than Weber's, and they all failed to find gravitational waves.

Weber published his results. Other scientists published theirs. Weber dismissed the negative findings of the other scientists. Experimental physicists studied Weber's results and said he was making mistakes. Perhaps the tape recorders he used to combine the data from the two cylinders were themselves accidentally injecting simultaneous signals. Or perhaps small magnetic fluctuations in electric power lines or lightning bolts could mimic gravitational waves. Weber held his ground. Theorists got in on the act. They calculated the amount of expansion and contraction that would be expected from realistic sources of gravitational waves in space. According to these calculations, Weber's resonant cylinders were not remotely sensitive enough to detect gravitational waves, even if such waves did indeed exist. Weber passionately held his ground. In telephone conversations, in personal visits, at scientific conferences, he got into scathing arguments. He lost friends and colleagues. Yet, in the face of a mountain of contradictory evidence, he continued to maintain that he was measuring gravitational waves. Clearly, Weber was not behaving in the traditions of science. Joseph Weber was allowing his personal investment to interfere with good judgment.

Then I, a greenhorn essayist, leaped into the fray. I wrote an essay about emotional prejudice in scientists for the magazine *Science '83*. The title: "Nothing but the Truth." In the essay, I ridiculed several scientists, including Weber. I cringe when I reread it. With self-righteous flourish, I wrote, "The white-haired Weber has become something of a tragic figure in the scientific community, continuing to declare his rightness in the face of incontrovertible evidence."

A few months after the essay was published, I found myself ten feet from Joseph Weber at a scientific conference. Some unsuspecting colleague introduced us. Weber's face immediately turned purple; he snarled something at me and stomped away.

Later, I decided that I deserved his contempt, and I hated myself for what I had written. Because Joseph Weber was really a hero. Yes, he was almost certainly sloppy in his experiment. And he should have graciously accepted the opposing results of other scientists. But he had imagined the first gravitational wave detector, he had built the first gravitational wave detector, and his insights about gravitational wave detectors had created the field. Today, the most advanced gravitational wave detector in the world, the Laser Interferometer Gravitational-Wave Observatory (abbreviated as LIGO), has just recently begun operations. If LIGO does not detect the first gravitational wave, then its successor probably will. LIGO would not exist without Weber's seminal work.

And it is quite possible that Weber would not have accomplished that work without his "emotional prejudice" and passion. In the book *Personal Knowledge,* the chemist Michael Polanyi argues that such personal passion is vital to the advance of science. I agree. Without a powerful emotional commitment, scientists could not summon up the enormous energy needed for pursuing an idea for years, working day and night in the lab or at their desks doing calculations, often sacrificing the rest of their lives. It is little wonder that such a personal commitment sometimes causes the scientist to defend his or her beliefs regardless of facts.

Even extraordinary physicists such as Einstein and Planck have defended their prejudices in the face of opposing evidence. Soon after Einstein published his theory of special relativity in 1905, a German experimental physicist named Walter Kaufmann repeated a crucial experiment to measure the mass of electrons moving at high speed. According to Einstein's theory, the mass of a moving particle should increase with speed in a particular way. A competing theory by Max Abraham, a colleague of Kaufmann's at Gottingen University, proposed a different formula for the increase in mass. Kaufmann's experimental results were closer to Abraham's predictions than to Einstein's. Over the next year, the great Max Planck, father of the quantum, carefully studied Kaufmann's experiment but could find no flaw. Nevertheless, Planck threw his support behind Einstein's theory.

Einstein himself, in a review article in 1907, said he could see nothing wrong with Kaufmann's experiments and agreed that they fit Abraham's theory better than his. Yet, he continues, "In my opinion [theories other than my own] have a rather small probability because their fundamental assumptions concerning the mass of moving electrons are not explainable in terms of theoretical systems which embrace a greater complex of phenomena." Here and elsewhere, Einstein clearly preferred his prejudice for comprehensive theoretical systems over actual experimental data. And data do sometimes change. A few years later, the experiments of Kaufmann were proven to be in error, and Einstein was vindicated. In future years, however, his prejudices sometimes led him astray. For decades, Einstein was personally committed to his nonquantum unified theory, which combined gravity and electromagnetism. In a letter to his friend Paul Erhenfest in 1929, Einstein wrote, "[My] latest results are so beautiful that I have every confidence in having found the natural field equations of such a variety." This time, Einstein turned out to be wrong. But that is not the point. Both when he was right and when he was wrong, Einstein's passion, his aesthetic and philosophical prejudices, and his personal commitment were probably essential to his scientific creativity.

All of which led me to question the meaning of the "scientific method." Since high school, I had been taught that scientists must wear sterile gloves at all times and remain detached from their work, that the distinguishing feature of science is the much-vaunted "scientific method," whereby hypotheses and theories are objectively tested against experiments. If the theory is contradicted by experiments, then it must be revised or discarded. If one experiment is contradicted by many other experiments, then it must be critically examined. Such an objective procedure would seem to leave little room for personal prejudice.

I have since come to understand that the situation is more complex. The scientific method does not derive from the actions or behavior of individual scientists. Individual scientists are not emotionally detached from their research. Rather, the scientific method draws its strength from the community of scientists, who are always eager to criticize and test one another's work. Every week, in many journal articles, at conferences, and during informal gatherings at the blackboard scientists analyze the latest ideas and results from all over the world. It is through this collective activity that objectivity emerges.

So how could I reconcile the Popperian view of science, with its unbudging demand for objective experimental testing, against the Polanyian view, with so much emphasis on the personal commitment and passions of individual scientists? The answer, perhaps obvious but at first shocking to a young scientist, is that one must distinguish between science and the *practice* of science. Science is an ideal, a conception of logical laws acting in the world and a set of tools for discovering those laws. By contrast, the practice of science is a human affair, complicated by all the bedraggled but marvelous psychology that makes us human.

ABOUT THE TIME of my ill-considered essay on Joseph Weber, I had a most beautiful experience with scientific discovery, perhaps the most beautiful of my life. I was studying the effects of particle creation in high-temperature gases. According to Einstein's famous formula $E=mc^2$, energy can be created from matter, and the reverse is also true. Matter can be created from energy. The phenomenon has been observed in the lab. It should also occur in space. Whenever the temperature of a gas is high enough, as should happen in strong gravity, then some of that thermal energy can be transformed into the creation of electrons and their anti-particles, the positrons. In turn, the creation of those particles will act back on the properties and emitted radiation of the gas. Thus, a good theoretical understanding of the nature of such a "relativistic thermal plasma" would be interesting not only in its own right, but also as a diagnostic for interpreting the gamma rays and X-rays observed from high-energy objects in space.

This research problem had been suggested to me by Martin Rees of the Institute of Astronomy, in England. I first met Martin during a visit to his institute in the summer of 1974, just after receiving my Ph.D. Martin was only thirty-two at the time. In the world of astrophysics, he was already a natural phenomenon himself. Among his many accomplishments, he was one of the first to point out that the distribution of quasars in space was inconsistent with the Steady-State theory of cosmology, thus lending support to the Big Bang theory. He has made major contributions to the astrophysics of black holes, the theory of galaxy formation, the origin of the cosmic background radiation, and many other topics. In fact, there has been practically no area of modern astronomy and cosmology that has not benefited from Martin Rees's imagination and original concepts. Martin is always erupting with new ideas, and he freely shares them without seeking acknowledgment or credit. Many of the nearly illegible letters I received from him during the middle and late 1970s, when we were working on similar problems, would begin "Thank you, Alan, for your very interesting preprint on X. I agree almost entirely with you, except for one or two small points." And then he would go on to elaborate on a number of important and often critical effects that I had missed in my investigation.

Many a pleasant summer I spent enjoying the unhurried pace and intimacy of Cambridge, England, walking around the luxurious gardens of the colleges and bicycling up the Madingley Road to the Institute of Astronomy. At that time, it was a modest one-story building, bordered by a wooden fence and a cow pasture. In the 1970s and 1980s, nearly everyone in the world worth their salt in astrophysics visited that building—to work quietly, to gather for British tea at four in the afternoon, and to catch ideas thrown out by the youthful but silver-haired Martin Rees, Plumian Professor of Astronomy and Experimental Philosophy. (In the 1990s, Martin became Sir Martin and was further elevated to Astronomer Royal of the United Kingdom.)

Sometime around 1980, Martin suggested the importance of understanding the theoretical properties of high-temperature gases. The problem nagged at me for a couple of years before I found a way to approach it. There were two obvious extreme cases. When temperatures were low, there would be no creation of particles. The properties of such a gas were well understood. In particular, the emitted radiation increased with increasing temperature in a known way. (All gases emit some radiation, except at zero temperature.) Also well understood was the case of extremely high temperatures. Here, there would be such a huge number of electrons and positrons created that the radiation would be trapped, except for a thin layer at the outer edge of the gas. The properties of this gas were well understood. In such a situation, the emerging radiation would have a well-known form, called "black-body radiation," also increasing with increasing temperature in a known way. However, because of the prodigious energy requirements, such extremely high-temperature gases with black-body radiation would not actually exist in space. Most interesting, therefore, is the intermediate case, when the temperature is high enough to create particles but not so high as to produce enough particles to trap the radiation and yield black-body radiation.

I was fascinated by the question of how the intermediate case would join the others. I expected that as energy was put into the gas at a higher and higher rate, the temperature would first start to increase according to the low-temperature case, then increase at some other intermediate rate, then finally increase according to the ultra-high-temperature case.

To my astonishment, I discovered something entirely different. With increasing energy input, the temperature at first did indeed rise as expected. But after increasing to a critical value, it began *decreasing* with further increase in the rate of energy input and emitted radiation. Finally, at a very high rate of energy input, the temperature turned around and began increasing again, in the known way for a very high-temperature gas.

At first this result seemed absolutely counter to my physical intuition. Put more energy into something and you expect its temperature to go up, not down. Then

I understood. The temperature of a gas is the average energy of a particle in that gas. Once you begin creating new particles, the additional particles can soak up all the increased energy, so much so that the average energy per particle can actually decrease. By analogy, when you give increasing quantities of food to a nation, the amount of food per person normally increases. But if the people of that nation produce children at a fast enough rate, then the food per person can actually begin decreasing, even though there is more and more total food.

The result was not only astonishing. It was delightful, it was beautiful, and it was a little mysterious. Again I experienced a kaleidoscope of emotions. Initially, I was surprised. Then, I was puzzled. Then, when I understood the result, I was extremely happy. I had found something new—again not terribly important in the grand scheme of science—but something that no one had ever known before me, and I felt elated and powerful with the knowledge. (In fact, a Swedish physicist, Roland Svensson, independently found the same result at about the same time, and we published nearly simultaneously.)

Then, I felt a sense of mystery. I had shed light on a small corner of nature. Other scientists had illuminated larger corners. But there were almost certainly vast chambers and ballrooms that remained in the dark. So many beautiful and strange things as yet unknown. In an article published in *Forum and Century* magazine in 1931, Einstein wrote, "The most beautiful experience we can have is the mysterious. It is the fundamental emotion which stands at the cradle of true art and true science." What did Einstein mean by "the mysterious"? I don't think he meant that science is full of unpredictable or unknowable or supernatural forces. I believe that he meant a sense of awe, a sense that there are things larger than us, that we do not have all the answers at this moment. A sense that we can stand right at the edge between known and unknown and gaze into that cavern and be exhilarated rather than frightened. Just as Einstein suggested, I have experienced that beautiful mystery both as a physicist and as a novelist. As a physicist, in the infinite mystery of physical nature. As a novelist, in the infinite mystery of human nature and the power of words to portray some of that mystery.

IN THE DECADE after my project on high-temperature gases, my science began gently subsiding, like a retreating blue tide. I looked out at the horizon and felt that my best work as a scientist was moving away into my past. At the same time, I gazed into the future and began pushing the boundaries of my essays. My essays took on more of a fabulist quality, like the writings of Italo

Calvino and Primo Levi. I invented. I told stories. I wrote an essay about life and society on a planet made entirely of iron. I wrote an essay about a moody Isaac Newton visiting my office. The science in my essays became only a doorway to what lay beyond. Eventually, when I was about forty years old, I began writing fiction. The time had arrived for my other passion to take over. By 1990, when I left Harvard for MIT, I had stopped doing scientific research altogether. I miss it terribly, despite the many pleasures and rewards of being a writer.

But I am still a scientist. I am still fascinated by how things work, by the beauty and logic of the natural world. When I see something interesting, like a particular angle made by the wake of a boat, I still take out a pencil and calculate why. When I travel on airplanes, I still amuse myself by re-deriving mathematical theorems that I learned years ago. Even when I write a scene for a novel, I sometimes subconsciously begin a paragraph with a topic sentence—a perfect metaphor for science but nearly fatal for art.

Every writer has a source for his writing, a deep, hidden well that he draws from to create. For me, I believe that source is science. In ways that I cannot explain, I believe that science suffuses all of my novels, characters, scenes, sentences, and even individual words. Some friends have told me that my novels have an architectural quality, a prominence of design. Perhaps that is a sign of the source.

Over the years, I have learned to recognize the different sensations of science and of art in my body. (Some of the sensations, such as the creative moment, are the same.) I know the feeling in my body of deriving an equation. I know the different feeling in my body of listening to one of my characters speak before I have told her what to say. I know the line. I know the swoop of a idea. I know the wavering note. Most of the time, these feelings all swirl together as a rumbling in my stomach, a wondrous and beautiful and finally mysterious cry of the world, logic and illogic, certainty and uncertainty, questions with answers and questions without.

(2003)

WORDS

AS A MEMBER of two communities, physicists and novelists, I've been fascinated by the different ways in which they work, the different ways in which they think, and their different approaches to truth.

One important distinction that can be made between physicists and novelists, and between the scientific and artistic communities in general, is in what I shall call "naming." Roughly speaking, the scientist tries to name things and the artist tries to avoid naming things.

To name a thing, one needs to have gathered it, distilled and purified it, attempted to identify it with clarity and precision. One puts a box around the thing and says what's in the box is the thing and what's not is not. Consider, for example, the word *electron*. As far as we know, all of the zillions of electrons in the universe are identical. There is only a single kind of electron. And to a modern physicist, the word *electron* represents a particular equation—the Dirac equation with field operators.

That equation summarizes, in precise mathematical

and quantitative terms, everything we know about electrons—every interaction, the precise deflections and twists of electrons by particular magnetic and electric fields, the tiny effects of electrons and their antiparticles materializing out of nothing and then disappearing again. In a real sense, the name *electron* refers to the Dirac equation. For scientists, it is a great comfort, a feeling of power, a sense of control, to be able to name things in this way.

The objects and concepts of the novelist cannot be named. The novelist might use the words *love* and *fear,* but these names do not summarize or convey much to the reader. For one thing, there are a thousand different kinds of love. There's the love you feel for a mother who writes to you every day during your first month away from home, and the love you feel for a mother who, when you stumble into the house drunk after driving home from the prom, slaps you and then embraces you. There's the love you feel for a man or a woman you've just made love to. There's the love you feel for a friend who calls to give you support after you've just split up with your spouse. But it's not just the many different kinds of love that prevent the novelist from truly naming the thing. It's also that the idea of love—the particular sensation out of the thousands of different kinds of love—must be shown to the reader not by naming it, but through the actions of the characters.

And if love is shown, rather than named, each reader will experience it and, what's more, will understand it in his or her own way. Each reader will draw on his or her own adventures and misadventures with love. Every electron is identical, but every love is different.

The novelist doesn't want to eliminate these differences, doesn't want to clarify and distill the meaning of love so that there is only a single meaning, like the Dirac equation, because no such distillation exists. And any attempt at such a distillation would undermine the authenticity of readers' reactions, destroying the delicate, participatory creative experience of a good reader reading a good book. In a sense, a novel is not complete until it has been read. And each reader completes the novel in a different way.

I'll give another illustration of the difference between naming and not naming. Let me represent science by expository writing. Like science, a piece of expository writing takes a reductionist and reasoned approach to the world. You have a position or argument, you structure this argument in logical steps, amassing facts and evidence to convince your reader of each assertion. We all learn that in expository writing it is good form to begin each paragraph with a topic sentence. A topic sentence, in effect, names the idea of the paragraph at the outset. You thus begin by telling your readers what they are going to learn in the paragraph and how to organize their thoughts so as to gain as ordered and structured an understanding as possible.

But in fiction, a topic sentence is usually fatal. Because the power of fiction is emotional and sensual. You want your reader to feel what you're saying, to smell it and hear it, to be part of the scene you are creating. You want your reader to be blindsided, to let go and be carried off to a magical place. Every reader will travel differently, depending on his or her own experiences of life. With a topic sentence, you don't leave room for your reader's own imagination and creativity. The difference can be stated in terms of the body. In expository writing, you want to go first to your reader's brain. In creative writing, you want to bypass the brain and go straight for the stomach, or the heart.

(2001)

METAPHOR IN SCIENCE

I REMEMBER THE DAY, during my first course in cosmology, when the professor was trying to explain how the universe could be expanding outward in all directions, but without any center of the expansion. To his credit, the teacher had covered the blackboard with equations, but we students still couldn't picture the situation. How could something explode uniformly in all directions without a middle of the explosion? Then, the professor said to pretend that space is two-dimensional and that the stars and galaxies are dots on the surface of an expanding balloon. From the point of view of any one dot, the other dots are moving away from it in all directions, yet no dot is the center. This powerful metaphor, first introduced by Arthur Eddington in 1931, has helped students of cosmology ever since, in every country and every language where the subject is taught. It works for anyone who has seen a balloon inflated.

Metaphor is critical to science. Metaphor in science serves not just as a pedagogical device, like the cosmic balloon, but also as an aid to scientific discovery. In doing science, even though words and equations are used with the intention of having precise meanings (as I have discussed in my essay "Words"), it is almost impossible not to reason by physical analogy, not to form mental pictures, not to imagine balls bouncing and pendulums swinging. Metaphor is part of the process of science. I will illustrate this point with some examples from physics.

In an essay on light and color in 1672, published in *Philosophical Transactions of the Royal Society,* Isaac Newton describes his first experiments with a prism. He darkened his chamber, made a small hole in his "window-shuts," and let a ray of sunlight enter a prism and spread out into colors on the opposite wall. He then interprets this phenomenon in terms of a theory of light:

> Then I began to suspect whether the rays, after their trajection through the prism, did not move in curved lines, and according to their more or less curvity tend to divers parts of the wall. And it increased my suspicion, when I remembered that I had often seen a tennis ball struck with an oblique racket describe such a curved line. . . . For the same reason, if the rays of light should possibly be globular bodies, and by their oblique passage out of one medium into another,

acquire a circulating motion, they ought to feel the greater resistance from the ambient ether on that side where the motions conspire, and hence be continually bowed to the other.

This passage is particularly revealing because it is a diary of Newton's personal thoughts in trying to understand the nature of light. Although Newton subsequently rejected the idea that light rays can curve through space, he continued to develop his corpuscular theory of light. In Query 29 of the *Opticks* (1704), Newton writes that rays of light are "bodies of different sizes, the least of which may take violet, the weakest and darkest of the colours and the most easily diverted by refracting surfaces." The largest and strongest light corpuscles carry red, the color least bent by a prism. Newton's mechanical worldview, of which light was only a part, held sway for two centuries.

A hundred years after Newton's *Opticks,* the physician and physicist Thomas Young allowed sunlight from his window to fall on a screen with two small holes in it. He then observed the alternating pattern of light and dark striking the opposite wall. From these patterns, Young proposed, in a paper entitled "Interference of Light" (1807), that light consisted of "waves" rather than particles:

Supposing the light of any given colour to consist of undulations of a given breadth, or of a given frequency, it follows that these undulations must be liable to those effects which we have already examined in the case of the waves of water, and the pulses of sound.

Young goes on to describe clearly the interference of two sets of circular waves moving outward from the two holes in his screen—a process of positive and negative reinforcement that would produce just the pattern of light seen on the wall. One cannot imagine how Young would have interpreted his observations without having seen overlapping ripples in a pond.

The great nineteenth-century physicist James Clerk Maxwell used an elaborate mechanical model, actually a sustained metaphor, to fathom the workings of electricity and magnetism. By analogy with fluids, Maxwell envisioned a magnetic field to be made out of closely spaced little whirlpools, which he called vortices. To solve the problem of what happens when two of these neighboring whirlpools touch and try to slow each other down, Maxwell mentally inserted between each pair of vortices a system of electric particles that would act as ball bearings. Designed for the purpose of reducing friction, the electric particles were also responsible for carrying electric current. The mechanical metaphors here are striking. In his paper "On Physical Lines of Force" (1861), Maxwell notes:

> In mechanism, when two wheels are intended to revolve in the same direction, a wheel is placed between them so as to be in gear with both, and this wheel is called an "idle wheel." The hypothesis about the vortices which I have to suggest is that a layer of particles, acting as idle wheels, is interposed between each vortex.

Maxwell's most celebrated contribution to the theory of electricity and magnetism was the prediction of oscillating waves of electric and magnetic forces, called electromagnetic waves, which travel through space at the speed of light. These waves were theoretically discovered by Maxwell after he added a single new mathematical term, called the "displacement current," to the equations of electricity and magnetism previously worked out by others. How did Maxwell deduce his hypothetical displacement current? When I first learned the theory of electricity and magnetism as a college physics major, I thought that the displacement current had been derived by requiring mathematical consistency and the conservation of electric charge. But, in fact, Maxwell was motivated by the demands of his mechanical model. He knew from experiments that an electric or magnetic field can store energy. Because Maxwell had a mechanical picture of his subject, and a mechanical view of the world, such energy could only be mechanical in nature. Maxwell therefore proposed that electrical and magnetic energy was stored by stretching, or displacing an elastic medium that filled up space, just as energy is stored in a stretched rubber band. In his classic paper of 1865, "A Dynamical Theory of the Electromagnetic Field," published in the *Royal Society Transactions,* Maxwell writes:

> Electric displacement, according to our theory, is a kind of elastic yielding to the action of the force, similar to that which takes place in structures and machines owing to the want of perfect rigidity of the connexions. . . . Energy may be stored in the field . . . by the action of electromotive force in producing electric displacement. . . . [I]t resides in the space surrounding the electrified and magnetic bodies . . . as the motion and strain of one and the same medium.

When an electric force oscillated in time, the electric particles in the elastic medium oscillated in response, giving rise to the so-called displacement current. The underlying elastic medium that allowed all of this to happen was called "ether." The ether was the material substance through which electromagnetic waves propagated, just as air is the substance through which sound waves propagate, by bumping one air molecule into the next. Again, in Maxwell's words, "We have therefore some reason to believe, from the phenomena of light and heat, that there is an aetheral medium filling space and permeating bodies, capable of being set in motion and of transmitting that motion from one part to another." Maxwell's mechanical model for electricity and magnetism, including the ether, was completely fictitious, but it led him to the correct equations.

In 1931, the Belgian priest and physicist Georges Lemaître published a stunning model for the origin of the universe. It was already known at this time—both from Lemaître's own theoretical work and from contemporary telescopic observations—that the universe is evolving. It was also known that very energetic particles, called cosmic rays, were constantly bombarding Earth from outer space, although the nature and origin of these particles was still a mystery. Lemaître proposed that cosmic rays originated billions of years ago from the radioactive disintegrations of enormous atoms and had been traveling through space ever since. Each of these ancient, massive atoms, in fact, was the parent of a star. Going back still further in time, when the cosmos was smaller and denser, the universe itself began as a single, giant atom, whose gradual disintegrations into smaller and smaller pieces formed nebulae, stars, and finally cosmic rays. In "L'expansion de l'espace," published in the November 1931 issue of *Revue des Questions Scientifiques,* Lemaître writes:

> We can conceive of space beginning with the primeval atom and the beginning of space being marked by the beginning of time. The first stages . . . consisted of a rapid expansion determined by the mass of the initial atom. . . . The atom-world was broken into fragments, each fragment into still smaller pieces. . . . The evolution of the world can be compared to a display of fireworks that has just ended.

In this example, both the literal and the metaphorical pieces of the metaphor arise from the unseen world of physics. Lemaître has been called the father of the Big Bang model of cosmology, but his primeval atom hypothesis went far beyond any theoretical or observational evidence.

The above examples should not be taken to mean that only the most brilliant scientists—the Newtons and the Maxwells—use metaphor in their work. Metaphor and analogy are rampant in physics. A graduate student and I recently calculated how gamma rays "reflect" from a layer of cold gas. In a discussion of this problem at the blackboard, beneath the equations, we drew a picture of a wavy line moving toward a wall. The line represented a single "particle" of light, called a photon, and the wall was the surface of the medium of gas. The conversation went something like this: "A high-energy photon penetrates very deeply into the medium before it first scatters off an electron. Then the photon scatters several more times, bouncing around, losing energy, and finally works its way back up to the surface of the medium, where it escapes."

I will end with an example from the forefront of theoretical physics—the string theory. The concept of strings has emerged from highly mathematical and formal attempts to describe the fundamental forces of nature. According to current string theory, the smallest unit of matter is not a pointlike object, but a one-dimensional structure called a "string." Here are some descriptions that leading string theorists have had the courage to put into print. "Scattering of strings is described by the simple picture of strings breaking and joining at the end." "Since a string has tension, it can vibrate much like an ordinary violin string. . . . In quantum mechanics, waves and particles are dual aspects of the same phenomenon, and so each vibrational mode corresponds to a particle."

Ultimately, we are forced to understand all scientific discoveries in terms of the items from daily life—spinning balls, waves in water, pendulums, weights on springs. We have no other choice. We cannot avoid forming mental pictures when we try to grasp the meaning of our equations, and how can we picture what we have not seen? As Einstein said in *The Meaning of Relativity,* "The universe of ideas is just as little independent of the nature of our experiences as clothes are of the form of the human body."

Sometimes different pictures of the same problem provide new insights. For example, the path that the Earth takes in orbiting the sun can be described either as a distant response to the sun itself, ninety-three million miles away, or as a local response to a gravitational field, filling or warping space around the sun. These two descriptions are mathematically equivalent, but they bring to mind very different pictures. As Richard Feynman comments in *The Character of Physical Law* (1965), they are equivalent scientifically but very different "psychologically," especially when we are trying to guess new laws of nature. In one picture, we focus on the two masses; in the other, on the space between them.

Physicists have a most ambivalent relationship with metaphor. We desperately want an intuitive sense of our subject, but we have also been trained not to trust too much in our intuition. We like the sturdy feel of the Earth under our feet, but we have been informed by our instruments that the planet is flying through space at a hundred thousand miles per hour. We find comfort in visualizing an electron as a tiny ball, but we have also been shocked to discover that a single electron can spread out in ripples, like a water wave, occupying several places at once. We crave the certainty of our equations, but we must give names to the symbols. At the age of twenty-five, Maxwell reflected on both the service and the danger of physical analogy in a paper entitled "On Faraday's Lines of Force" (1856), published in the *Transactions of the Cambridge Philosophical Society:*

> The first process therefore in the effectual study of science, must be one of simplification and reduction of the results . . . to a form in which the mind can grasp them. . . . We must, therefore, discover some method of investigation which allows the mind at every step to lay hold of a clear physical conception, without being committed to any theory founded on the physical science from which that conception is borrowed.

When the quantum theory was being developed, in the first two decades of this century, physicists agonized over their inability to picture the wave-particle split personality of subatomic particles. In fact, physicists violently disagreed over whether such pictures were even useful. In 1913, Niels Bohr, a pioneer of quantum theory, proposed a model for the atom in which electrons orbited about a central nucleus. Max Born commented a decade later (*Die Naturwissenschaften,* vol. 27) that "a remarkable and alluring result of Bohr's atomic theory is the demonstration that the atom is a small planetary system . . . the thought that the laws of the macrocosmos in the small reflect the terrestrial world obviously exercises a great magic on mankind's mind." Referring to this same Bohr model, Werner Heisenberg warned that "quantum mechanics has above all to free itself from these intuitive pictures . . . that in principle [are] not testable and thereby could lead to internal contradictions" (*Die Naturwissenschaften,* vol. 14).

In the 1920s and 1930s, there were two competing formulations of quantum theory, eventually shown to be mathematically equivalent. Heisenberg was the architect of the highly abstract version; Erwin Schrödinger had worked out a more visual theory. In a letter to Wolfgang Pauli in 1926, Heisenberg wrote, "The more I reflect on the physical portion of Schrödinger's theory the more disgusting I find it."

Schrödinger, in his reply in the *Annalen der Physik,* wrote that he felt "repelled by the methods of transcendental algebra" in Heisenberg's theory, "which appeared very difficult" and had a "lack of visualizability." A few years later, in 1932, Professor Heisenberg ascribed the nuclear force between a proton and a neutron to the "migration" (*Platzwechsel*) of an electron between them. Where did the alphas and betas in Heisenberg's matrices or the readings from the electrometers say that an electron could migrate? Remarkably, Heisenberg's image of migrating particles as agents of force led three years later to Hideki Yukawa's successful prediction of a new elementary particle, the meson.

Ultimately, Bohr himself was frustrated in his attempts to grasp intuitively the world of the atom. In 1928, he lamented (*Nature Supplement*, April 14, 1928):

> We find ourselves here on the very path taken by Einstein of adapting our modes of perception borrowed from the sensations to the gradually deepening knowledge of the laws of nature. The hindrances met with on this path originate above all in the fact that . . . every word in the language refers to our ordinary perceptions.

Many contemporary physicists have essentially given up trying to describe the fundamental elements of nature by anything based on common sense. Richard Feynman has remarked that he can picture invisible angels but not light waves. Steven Weinberg, like Bishop Berkeley, seems on the brink of abandoning the material world altogether when he says that after you have described how an elementary particle behaves under various mathematical operations, "then you've said everything there is to say about the particle . . . the particle is nothing else but a representation of its symmetry group" (R. Crease and C. Mann, *The Second Creation: Makers of Revolution in Twentieth-Century Physics*, 1986). Yet physicists still use metaphors. Cosmologists still discuss how the universe "expanded and cooled" during the first nanosecond after its birth. Relativists still talk about the "semipermeable membrane" around a black hole. String theorists still describe their unseen subatomic strings as "stretching, vibrating, breaking." What other choice do we have? We must breathe, even in thin air.

But there is a difference between metaphors used inside and outside of science. In every metaphor, there is a principal and a subsidiary object, the literal and the metaphorical, the original and the model. When we use metaphors in ordinary human affairs, we usually have a good sense of the principal object to begin with. The metaphor deepens our insight. When we hear that "the chairman plowed through the discussion," we already know a good deal about chairmen, committees, and tiring discussion. But when we say that a photon scattered off an electron, what concrete experience do we have with electrons or photons? When we say that the universe is shaped like the surface of a balloon, what do we really know about how space curves in three dimensions?

Galileo admired Copernicus for being able to imagine

that the Earth moved, against all common sense. But at least Copernicus understood that the Earth was a ball and had seen other balls move. The objects of physics today, by contrast, are principally known as runes in equations or blips from our instruments. Earlier physicists had an immediate and tactile relation to their subjects. Descartes could see the disjointed image of a pen half in water and half in air. Du Fay could rub cats' fur against copal or gum-lack or silk. Count Rumford could feel the heat in a cannon just bored. But in the last century or so, science has changed. Physics has galloped off into territories where our bodies cannot follow. We have built enormous machines to dissect the insides of atoms. We have erected telescopes that peer out to unimaginable distances. We have designed cameras that see colors invisible to human eyes. Theorists have worked out equations to describe the beginning of time. The objects of physics today are far removed from human sensory experience.

As a result, it seems to me that metaphors in modern science carry a greater burden than metaphors in literature or history or art. Metaphors in modern science must do more than color their principal objects; they must build their reality from scratch. Such substance, in the palm of a modern physicist, is often hard to let go of. Although aware that the ether was based only on mechanical analogy, Maxwell believed it existed. The year before he died, Maxwell wrote in the ninth edition of *The Encyclopaedia Britannica* (1878) that he had "no doubt that the interplanetary . . . spaces are not empty but are occupied by a material substance or body, which is certainly the largest . . . body of which we have any knowledge." If a giant of science like Maxwell was seduced by his own metaphor, what can happen to the rest of us? We ought not to forget that when physicists say a photon scattered from an electron, they are discussing that which cannot be discussed. We can see the tracks in a cloud chamber, but we cannot see an electron. Metaphors in science—although a critical part of our reasoning and discovery—should be handled with caution, and with a clear knowledge of the limits of our sensory experience of the world. We are blind people, imagining what we don't see.

(1988)

INVENTIONS OF THE MIND

MATHEMATICS, FOR ME, has always been a means to an end. Like many other scientists, I have some facility with math. But I found from a young age that I could not become passionate about a problem unless it had physical meaning. How did a radio work? Why did a spoon halfway in water appear to bend sharply in two? Why was upstairs usually warmer than downstairs? Growing up, I stocked a closet off my room with capacitors and resistors, wire and batteries, test tubes and flasks. I did experiments. When I took algebra, I loved the x's and the y's, their purity and their power, but I knew that they stood for eggs or coins or the ages of children. Later, I had friends who were mathematicians. They were not interested in the eggs or the coins. For them, the x's and the y's were enough. For them, the x's and the y's lived in a world of their own.

Yet despite my preferences for reality, I've always been haunted by the conviction of Einstein, a physicist, that the deep truths of nature cannot be uncovered by experiment. Rather we discover the deep truths of nature, he says, only by "free inventions of the mind." And by mathematics, in particular. In a lecture delivered to Oxford in 1933, Einstein said, "I am convinced that we can discover by means of purely mathematical constructions . . . the key to the understanding of natural phenomena. Experience may suggest the appropriate mathematical concepts, but they most certainly cannot be deduced from it. . . . The creative principle resides in mathematics. In a certain sense, therefore, I hold it true that pure thought can grasp reality, as the ancients dreamed."

How could that be? Surely we don't invent the material world around us. But it often seems that we do. Or rather, it seems that mind often comes first, with matter catching up to it later. Leucippus conceived of the atom twenty-five centuries before its discovery. Paul Dirac predicted the existence of antimatter in 1930, based on his highly mathematical equation for the electron; the first antiparticle, the positron, was found two years later. In 1864 James Clerk Maxwell used his equations for electricity and magnetism to predict the possibility of radio waves and other traveling undulations of pure energy; radio waves were first produced and detected twenty years later. How does that happen? How can pure thought have such power to predict?

I WAS FULL of such questions one morning in the fall of 1993 when I visited the mathematics department at Princeton, a citadel of pure mathematical thought. Princeton has always been the home of great mathematicians, such as forty-one-year-old Andrew Wiles, who had just proven Fermat's Last Theorem. In fact, I knew the Princeton campus well. I'd been an undergraduate there many years before, a physics major. But I'd rarely set foot inside Fine Hall, the mathematics building, a smooth narrow tower of concrete and glass.

I began idly wandering through the building. As I expected, the cluttered offices and the blackboards brimmed with strange notations, representing sets, elliptic functions, infinite series, convergences, spheres in eight dimensions. There were no references to the Hubble telescope or to AIDS or the postmodern novel or the end of the cold war. By and large, the mathematicians at Princeton, and everywhere else, work in pure ideas. They imagine worlds devoid of bodies, of whim, of physical substance of any kind. For many, the powdery chalk on their blackboards is all they want of material reality. As it was put by the great British number theorist G. H. Hardy, "Imaginary universes are so much more beautiful than this stupidly constructed 'real' one." Still, the voice of Einstein whispered in my ears: the imagined universe somehow congeals into the real one, as stupid as it may be.

MATHEMATICS MIGHT BE divided into two categories, applied mathematics and pure mathematics. Applied mathematics is propelled by an understanding of the physical world. The Babylonians and Sumerians developed geometry, among other reasons, to measure the area of fields. Newton invented calculus to aid his studies of motions, the velocities and accelerations of pendulums and planets. Televisions, computers, and advice on trips to Las Vegas all use applied mathematics. Pure mathematics has no such motivation. Pure mathematics exists within a world of its own, a mathematical world. It lives for itself. Pure mathematics is pure mind, the weightless domain where my questions began.

Irrational numbers, for example. No one knows when numbers first appeared. In some ancient cave, perhaps, in firelight, men and women scratched marks in dust, a separate stroke for each finger on the human hand. Perhaps these early people counted berries, deer and bison, risings of the sun between full moons, windings of streams and rivers. Children counted pebbles. Numbers were counted things, were measured things. Numbers could be seen or touched. A length of twine could gauge the distance around a rock, a field of grass. The twine could be folded into halves or thirds or eighths to measure fractions of itself. A fist might be ⅙ a length of twine, a stone might be ⅞. There were the whole numbers and the fractions, both applied mathematics.

Thousands of years later, around 450 B.C., the Greeks made an astounding discovery. They found numbers that were neither whole nor fraction. For centuries, human beings had reasonably thought that if a ruler had fine enough gradations, it could measure any length exactly. The Greeks, however, found that some lengths can never be gauged as fractions of a foot, or fractions of any unit of length, no matter how minuscule the gradations. For example, take a square whose side is one inch long. You cannot exactly measure the square's diagonal with a ruler. Divide the ruler, divide it again and again, continuing to an *infinite* number of divisions with an *infinitely small* space between them, and the end of the diagonal will always fall between two marks. The length of the diagonal is $\sqrt{2}$ inches, where $\sqrt{2}$ is the number that when multiplied by itself gives exactly 2. The number $\sqrt{2}$ is a new kind of number, called an irrational number. (Irrational numbers, like $\sqrt{2}$, cannot be written as the ratio of two whole numbers; all other numbers, such as $0.4 = ⅖$, can be expressed as the ratio of whole numbers and hence are called rational numbers.) Irrational numbers are pure mathematics. For any practical purpose, for any actual measurement, the rational numbers work admirably well because the length of any timber or stone is infinitely close to some rational number. But the Greeks were not content with practical purposes. By thought alone, they found numbers that cannot be measured.

SOMETHING OF THE power of pure mathematics can be seen by its permanence. We often use the terms *objective* and *subjective* to distinguish between those things existing outside our minds versus those produced in our minds, things with reality and permanence versus things that shift and dissolve with each new point of view. Paradoxically, mathematical results, which we deduce in our minds, last forever. Gone is the civilization of ancient Greece, but not the Pythagorean theorem, which states that the square of the hypotenuse of a right triangle equals the sum of the squares of its two legs. Euclid's theorem that the angles of a triangle sum to 180 degrees will *always* be true. New branches of mathematics have been developed, including new branches of geometry, but the angles of a Euclidean triangle will always sum up to 180 degrees.

Science, by contrast, is constantly revising itself, constantly changing its theories and results to give better approximations to physical reality. An example is the theory of light. The wave nature of light was first revealed in the mid-seventeenth century by the experiments of the Italian scientist Francesco Maria Grimaldi. Grimaldi discovered that the concentric circles of darkness and light produced by light emerging from a small hole are similar to the crests and the troughs of overlapping waves of water. In the mid-nineteenth century came Maxwell's fully quantitative theory of light as an electromagnetic wave. Maxwell's equations, however, did not correctly explain all phenomena of light. In the early twentieth century, experiments suggested that light does not always behave as a continuous wave; it sometimes acts as a group of discrete particles, or "quanta," called photons. In the late 1940s, physicists produced a new quantitative theory of light called quantum electrodynamics, replacing Maxwell's equations. In the 1960s, scientists went even further and proposed that the phenomenon of light is deeply connected to other fundamental forces. The theory of light was modified once again. Scientists have differing opinions about whether humankind will ever find the ultimate laws of nature, or whether such final truths exist. But there is no disagreement that the history of science is the history of an endeavor constantly revising and refining its laws. Scientific theories do not have the staying power of pure mathematics.

EINSTEIN'S "free inventions of the mind" refers to the mind, the human mind. And inventions. When a mathematician is working on a problem, his or her unconscious mind is constantly considering many possibilities. Why are some brought to the level of consciousness and others left in the murky depths? Why do some ideas feel so compelling? The French mathematician Henri Poincaré addressed this question in a lecture delivered to the Psychological Society of Paris at the turn of the century:

> The privileged unconscious phenomena, those susceptible of becoming conscious, are those which directly or indirectly affect most profoundly our emotional sensibility. It may be surprising to see emotional sensibility invoked *à propos* of mathematical demonstrations which, it would seem, can interest only the intellect. This would be to forget the feeling of mathematical beauty, of the harmony of numbers and forms, of geometric elegance. This is a true aesthetic experience that all real mathematicians know, and it surely belongs to emotional sensibility.

The aesthetics of the human mathematical mind, as well as the power and finality of pure mathematics, are no better conveyed than in the proof of a theorem. A mathematical theorem is a statement about the nature of mathematical reality, about the behaviors of numbers and curves in the intangible universe of mathematics. The proof of a theorem is the logical argument that shows the theorem to be true. Mathematicians are interested in much more than the truth or falsity of a conjecture. A "good" proof also often introduces new concepts and techniques. And it should reduce the conjecture to a self-evident statement, one that we understand clearly and deeply, practically a tautology.

AN ELEGANT EXAMPLE is a theorem about prime numbers, posed and proved around 300 B.C. by the Greek mathematician Euclid. A prime number is a whole number that is not evenly divisible by any other whole number. For example, 17 is a prime number, but 36 is not. Prime numbers are fundamental to the theory of numbers. One can regard the prime numbers as the building blocks out of which all other numbers are built. Any number can be written as a product of primes, simply by dividing its factors into smaller and smaller numbers until each factor can be divided no longer. For example, we can write the number 530 as 530 =

$2 \times 265 = 2 \times 5 \times 53$, where the last three numbers are prime.

One might plausibly think that there is a largest prime number, a largest number that cannot be divided by other numbers. Any larger number would be divisible by one of the huge supply of smaller numbers. Euclid's theorem says no: There is no such largest prime. The primes continue indefinitely, to infinity.

Euclid's proof of his theorem uses a technique called *reductio ad absurdum*. In this method, a result is proven true by showing that its falsity would lead to a logical contradiction. And there are no contradictions in mathematics. Here is Euclid's proof, which I first learned in high school:

Suppose there were a largest prime. Call it P. Then the complete list of primes, in ascending order, would be 2, 3, 5 . . . P, where the . . . stands for all the intermediate primes between 5 and P. Now consider the number $Q = (2 \times 3 \times 5 \times \ldots P) + 1$, that is, the product of all primes up to P, plus 1. The number Q is certainly larger than P and therefore not prime, by our starting supposition. Thus Q must be divisible by a number, and since all numbers are the products of primes, Q must be divisible by a prime. By its definition, Q is not divisible by any of the primes up to P, because dividing it by any of these numbers leaves a remainder of 1. Therefore it must be divisible by a prime larger than *P*. But this conclusion contradicts our starting hypothesis that there are no primes larger than *P*. So that hypothesis must be false. There is no largest prime. The proof is finished. Or, as mathematicians love to say, QED, which is short for the Latin phrase *quod erat demonstrandum,* "which was to be proved."

Mathematical proofs are not only about mathematics. They are about mathematicians as well. Euclid's proof is revealing for what is absent, as well as for what is present. There are no unnecessary elements, no false starts, no wrong turns. Undoubtedly, the first attempted proofs of the conjecture were clumsy, and Euclid must have thrown out a lot of good papyrus before he produced the proof that he handed down to the ages. Indeed, the ideal proof in mathematics shows no traces of the mortal path of trial and error that led to it. Johann Carl Friedrich Gauss, the great German mathematician, often called the "Prince of Mathematics," refused to publish his mathematical proofs until he had fashioned them into works of disembodied perfection. A cathedral is not a cathedral, he said, until the last scaffolding is down. When one looks at a proof by Gauss, it is impossible to tell where his reasoning began. Gauss wanted it that way. Pure mathematics is often compared to an art form, but it is a peculiar art form. A mathematical proof is a beautiful painting in which the viewer is not supposed to see the brushstrokes of the artist. That absence and simplicity is part of the aesthetic.

I have no doubt that Einstein, and other great scientists, have also been motivated by aesthetics in their search for new laws of nature. The Nobel physicist Steven Weinberg, who proposed the successful unified theory of the electromagnetic and weak nuclear forces far ahead of experimental evidence, uses words like *beautiful* and *ugly* to describe theories he likes and doesn't like. Dirac, whose equation for the electron has widely been described as one of the most "elegant" and "beautiful" in all of physics, used to say "Find the mathematics first, and think what it means afterwards." The Nobel chemist Roald Hoffmann tells his students that it is the awareness and appreciation of the "aesthetic aspects of science," rather than mere quantitative analysis, that leads to discovery. And Einstein himself, in his *Autobiographical Notes,* says that the selection of physical theories has always been guided by considerations of "naturalness" and "logical simplicity" and "inner perfection." Perhaps aesthetic criteria are part of Einstein's "free inventions" of the mind.

I WANTED TO LOOK OUT. The best view from Fine Hall is on the top floor, the twelfth, high above the other buildings on campus. From one window, I could gaze down on the twin domes atop the astronomy building, Peyton Hall. From another, I could see Jadwin Hall, the physics building, also far below. Literature, philosophy, foreign languages sat demurely on the other side of Washington Road. Surrounding houses of the town, tiny from that altitude, scattered out and merged into the leafy countryside, the encircling trees, yellow and orange and red in October, the colors slowly fading in the distance.

About four o'clock, I went down to tea. Every afternoon, the mathematicians in Fine gather on the second floor for tea. At the back of the room loomed a large photograph, a conference of great mathematicians from the 1940s. They were lined up in rows, staring off into space.

One might think that, living in their beautiful worlds of sublime isolation and perfection, mathematicians would be the happiest of all people. However, many don't seem at peace with their chosen profession. Mathematicians are ruthlessly self-critical. In most professions, it is possible to tell yourself and others that your accomplishments are significant, whether they are or are not. Not so in mathematics. In the community of mathematicians, there is a disturbing consensus on what is important, and the standards are painfully high. "Mathematicians are more aware of their failures than any other professionals," says Professor Gian-Carlo Rota of MIT. Of his own work, Rota says that only one or two moments have brought him any pleasure. Looking back on his long career, Hermann Weyl, the eminent mathematician at the Institute for Advanced Study, told a colleague that he considered his life a failure. Near the tearoom of Fine, I ran into Simon Kochen, the past chairman of the Princeton mathematics department. Kochen, a trim and articulate man, leaned in a doorway and said that "the moments of joy in mathematics are few and far between. Most of math is pure frustration. Results, when you finally get them, are obvious." (Isn't that the goal of a good proof, anyway, to reduce the proposition to a near tautology?) Many mathematicians keep most of their calculations permanently in file drawers, having decided that their results are not worth publishing.

TEA WAS OVER and the day coming to an end. I wanted to get back to my questions. The great irony and mystery of mathematics, and the ultimate example of matter succumbing to mind, is that pure mathematics often becomes applied mathematics. That is, purely mathematical ideas emanating from the minds of mathematicians, with no physical meaning in sight, often later become essential to understanding the material world.

Imaginary numbers, for example. When ordinary numbers, positive or negative, are multiplied by themselves, the result is a positive number. A different type of number was conceived in the sixteenth century. The "imaginary number" called i is the number that when multiplied by itself gives -1. Imaginary numbers, although useful to mathematicians, remained imaginary for centuries. Today, physicists cannot do without them. Imaginary numbers are used to describe how colliding waves interfere with each other. Imaginary numbers enter the calculations of the swing of a pendulum in air. The fundamental equation of quantum physics, the Schrödinger equation, explicitly contains an imaginary number.

Another striking example is non-Euclidean geometry. Euclidean geometry, laid out systematically by Euclid over two thousand years ago, is the geometry we learn in school. That the angles in any triangle add up to 180 degrees is a result of Euclidean geometry. That the circumference of any circle divided by its radius is a universal number, 2π, is a result of Euclidean geometry. Euclidean geometry accords with our physical sense of the world. In the nineteenth century, Friedrich Riemann and other mathematicians developed geometries that were different from Euclid's but still mathematically self-consistent. These geometries correspond to curved spaces, like the surface of a beach ball. In non-

Euclidean geometry, the angles of a triangle do not sum up to 180 degrees; a circle's circumference divided by its radius is not equal to 2π, and so on. Non-Euclidean geometries were pure mathematics.

Fifty years later, in 1912, Einstein found that non-Euclidean geometry was just what he needed for his new theory of gravity, called general relativity. In general relativity, the effect of gravity is equivalent to a modification of the shape of space, a warping of space. In this picture, a gravitating body like the sun distorts space around it into the shape of a funnel, with the sun at the center. Orbiting planets travel along the surface of the funnel. The quantitative formulation of general relativity uses the rarefied mathematics of non-Euclidean geometry. Subsequent experiments have confirmed many of the physical predictions of general relativity. But Riemann didn't have physical space in mind. His calculations, in fact, refer to abstract, mathematical spaces. Riemann was a pure mathematician.

WHY DOES IT HAPPEN that pure mathematics so often finds application to nature?

The mundane explanation is that it is our human description of natural phenomena, rather than the phenomena themselves, that rely on mathematics. In this view, mathematics is just another human language. And when we wish to describe a new phenomenon of nature, we draw on whatever vocabulary is at our disposal. The subatomic world, for example, does not *necessarily* have to be described with imaginary numbers; a description of gravity does not *necessarily* have to employ non-Euclidean geometry, or any geometry at all, and so on. Schrödinger employed imaginary numbers in his quantum equation and Einstein enlisted non-Euclidean geometry in his general relativity merely because those methods and concepts were in their vocabularies and fit. Other methods might have been employed, just as a scene might be described in different words by two different writers. In this view, every idea in pure mathematics will eventually get "used" for the same reason that every word in a language eventually winds up in a sentence. One might consider this analysis a literary explanation of applied mathematics.

The literary explanation could be right. I don't know whether quantum physics could be reformulated without imaginary numbers, but I do believe that such a reformulation, if possible, would appear enormously awkward compared with the current version. The literary explanation also fails to confront the extraordinary success of mathematics in describing the physical world, a success that has impressed nearly everyone who has worked professionally in science. The equations of quantum electrodynamics, after long mathematical manipulations, predict that the electron should have a magnetic strength of 1.00115965214. The value measured in the lab is 1.00115965219. Even if the equations could be restated in words, it is hard to imagine how those words could be manipulated to produce such an astonishing result. In his little book *The Character of Physical Law*, the physicist Richard Feynman writes about mathematics: "If you want to learn about nature . . . it is necessary to learn the language that she speaks in. She offers her information only in one form."

A MORE INTERESTING explanation of why pure mathematics often finds physical application is that all science is a human construction. This position, sometimes associated with the new British school of the philosophy of science, is part of a recent trend in many intellectual disciplines to overthrow absolute meaning and to look instead for the cultural context of works and ideas. Under this interpretation, since science is a human activity, its laws and results are constructed by humans. We might call this notion the constructionist explanation of applied mathematics. It reduces my original question to a tautology. Pure mathematics and science *both* lie within the human mind. Therefore, it is no wonder that they should relate to each other, just as two successive thoughts often relate to each other. (The notion that the physical world might lie within the mind of the observer does, of course, have a long philosophical tradition and is especially associated with George Berkeley.)

There is no doubt that the activity of scientists and even many of their ideas are dependent on social and cultural factors. But it is extremely hard to deny the existence of something real out there independent of our minds. The mass of the proton is 1,836 times that of the electron. Snowflakes have a six-sided symmetry. How could we imagine these things? If the physical world were just a reflection of ourselves, then there would never be surprises in science. But we are constantly surprised. The experimental discovery in the late nineteenth century that the propagation of light does not require a material substance, the so-called ether, so contradicted expectations and plain common sense that the principal researcher, Albert Michelson, believed his experiment had failed. He kept repeating it for years. (For his "failure," Michelson was awarded the Nobel Prize in 1907, the first American to win a Nobel in the sciences.)

A third explanation of the paradox, and also I think an extreme one, is that our minds are part of nature, tuned in somehow in a fundamental way. In contrast to the previous explanation, the physical world indeed has its own independent existence, but somehow, perhaps through the eons of evolution, our human brains have acquired the inherent logic of nature. If our minds, in some sense, reflect nature, if our logic is nature's logic, then the eventual application of pure mathematics does not seem surprising. One might consider this hypothesis a holistic explanation of applied mathematics. The holistic explanation, similar to the old notion of animism, may be associated with other recent ideas such as the "Gaia hypothesis" and the "anthropic principle." The Gaia hypothesis, developed by biologists, views the Earth as a single living organism, its various living and nonliving systems all working together in harmonious feedback loops. The anthropic principle, discussed by physicists, suggests that the physical properties of our universe are exquisitely tuned to allow the existence of life. Slightly different properties of the cosmos, such as a weaker nuclear force, would not permit the emergence of life or of intelligence to contemplate that cosmos.

FINALLY, THERE IS what one might call the mathematical explanation of applied mathematics. Perhaps the physical world is, underneath, a giant mathematical system. Maybe Descartes and Newton and their followers didn't go far enough in conceiving of the universe as a giant clock. Perhaps the material universe is pure mathematics, an extreme version of Plato's ideal forms. In this view, the physical objects we see, such as electrons or galaxies, are just material manifestations of an underlying mathematical plan, just as the many kinds of chess-boards and arrangements of pieces on those boards are just representations of the abstract rules of the game. Perhaps there is no way the universe could have been put together except by following pure mathematics. In this interpretation, mathematical reality and physical reality, although seemingly different, are really part of a single reality. It would be no wonder, then, that pure mathematics often finds application to the physical world; it was there to begin with.

Even in a mathematical world, however, experiments could not be relied on to uncover the truths of nature because experiments are always imprecise, and they are limited in scope. Real experiments always have unwanted disturbances, called "noise," that obscure the pure voice underneath. A tiny bit of air always leaks into a vacuum. A tiny bit of stray light always confuses a photodetector. And real experiments, like our human ears and eyes, are limited to certain frequencies and sizes and sensitivities.

But mental experiments have no such limitations. Mental experiments have the purity of the mathematics underneath. Returning to Einstein's approach to science, we might do the mental experiments first, then test against nature, finding what parts of the world we have just found in our minds.

Both the holistic and the mathematical explanation of

applied mathematics are beautiful but mysterious, and they certainly leave many unanswered questions. However, the fact remains that both scientists and mathematicians alike have been utterly baffled by the powerful connection between pure mathematics and the physical world. As I left Princeton's Fine Hall for the day, I felt as perplexed, and amazed, as when I had entered that morning.

In *The Heart of Darkness*, Conrad writes that "the mind is capable of anything—because everything is in it, all the past as well as all the future." It is pleasing to think that somewhere in our minds perhaps lies a building waiting to be built, a grand unified theory of physics, the beautiful song of a hermit thrush, a sentence waiting to be written.

(1994)

THE CONTRADICTORY GENIUS

ONE

UNTIL THE AGE of seven or eight, whenever the young Albert Einstein was asked a question, he would slowly formulate an answer, mutter it tentatively to himself, and finally repeat aloud his considered response. This laborious method of speaking gave the impression that he needed to say everything twice. His parents consulted a doctor, and the family housekeeper called the boy "stupid." Decades later, Einstein's sister, Maja, recorded this odd childhood habit and attributed it to her brother's thoroughness in thinking. Yet the doubling of each sentence, once for himself and once for everyone else, may also have been an early sign of the deep inner world that Einstein inhabited. Brilliant, supremely self-confident, brutally honest, witty, stubborn—Einstein was above all else a loner.

In an essay he published in *Forum and Century* in 1931, at the age of fifty-two, the physicist wrote:

My passionate sense of social justice and social responsibility has always contrasted oddly with my pronounced lack of need for direct contact with other human beings and human communities. I am truly a "lone traveler" and have never belonged to my country, my home, my friends, or even my immediate family, with my whole heart; in the face of all these ties, I have never lost a sense of distance and a need for solitude—

With this passage one cannot help recalling a close contemporary of Einstein's, the poet Rainer Maria Rilke, who famously advised another young poet to "love your solitude and bear with sweet-sounding lamentation the suffering it causes you." Einstein's isolation was surely in part the artist's compulsion to create alone. And the great physicist was indeed an artist in his devotion to simplicity and mathematical beauty. But his distance went far beyond any aesthetic concerns. Throughout his life, he maintained a strong awareness that he did not fit in, intellectually, socially, spiritually. Einstein had a profound sense of otherness, even alienation.

Numerous anecdotes from childhood suggest that these feelings were partly a consequence of innate temperament. But they were also strongly accentuated by his harsh and authoritarian early teachers, the German military service that caused him to renounce his citizenship at the age of sixteen, his parents' contempt for his sweetheart and first-wife-to-be, Mileva Marić, his inability to secure university employment after college, and, finally, his growing identification with the plight of his fellow Jews, whom he referred to as his "tribal companions."

The 1931 passage on solitude, together with many of Einstein's other public essays, have long been available in two of his books, *Mein Weltbild* (1934) and *Ideas and Opinions* (1954). (The latter is an extraordinary compendium of Einstein's thoughts on philosophy, religion, education, politics, and the methods of science.) These writings, from later life and after his rise to worldwide fame, conform more or less to the popular image of Einstein as a wise, grandfatherly figure. When hints of the lone traveler appear here and there, they are couched in rather abstract language. A more private and gritty view of Einstein is now emerging in several new biographies, stimulated by the recent availability of the younger Einstein's personal letters and perhaps also by the recurrent fascination with the frailties of our heroes. Beyond the usual revealed intimacies and imperfections is clear evidence that Einstein's sense of estrangement began at a young age. Of the new biographies, the fullest is Albrecht Fölsing's huge *Albert Einstein*, originally published in German in 1993 and now appearing in English.

PRIME AMONG the new source material is a set of fifty-two love letters exchanged between Einstein and Mileva Marić, extending from 1897 to 1902. These letters were released to scholars only in 1986, upon the death of Einstein's older son, Hans Albert. They have been printed in full in the first volume of *The Collected Papers of Albert Einstein,* an immense project jointly sponsored by the Hebrew University of Jerusalem and Princeton University Press. (At his death in 1955, Einstein willed all of his papers and letters to the Hebrew University; the Einstein Archive there now contains some forty-five thousand documents, and *The Collected Papers* is expected to run to over twenty-five volumes.)

Mileva Marić, born in 1875 and three years older than Albert, came from a well-to-do Serbian family living in what was then Hungary. Darkly pretty, with a slight limp from childhood, she excelled in physics and mathematics. At the time that her surviving correspondence with Albert began, in October 1897, they were both in their second year of study at the Polytechnic in Zurich—Switzerland being the only German-speaking country that then allowed women into higher education. Mileva had just taken a temporary leave to attend lectures at the University of Heidelberg.

From these letters we learn, in Einstein's own words, how he is tormented by his parents over Marić. In 1900, he writes her: "My parents are very distressed about my love for you, Mama often cries bitterly & I am not given a single undisturbed moment here. My parents mourn for me almost as if I had died." A hint of Mama's objections is given in a letter Einstein wrote to Marić a few months earlier, in which he quotes his mother as saying: "She is a book like you—but you ought to have a wife." "When you'll be 30, she'll be an old hag."

Other letters convey a strong sense that the young physicist was battling the world, not just his parents. In 1901, Marić failed her degree examination at the Polytechnic and moved back with her parents while she was pregnant with Einstein's illegitimate child. (The child, a daughter named Lieserl, was evidently given up soon after birth; no records have been found, and she has vanished from history.) Einstein, meanwhile, was living close to poverty and having no luck in obtaining a job. It is under these circumstances that he writes to her:

I decided the following about our future; I will look *immediately* for a position; no matter how humble. . . . The moment I have obtained such a position I'll marry you and take you to me without writing anyone a single word before everything has been settled. And then nobody can cast a stone upon your dear head, and whoever dares to do anything against you, he'll better watch out!

Then, in December 1901, just after Einstein has learned that he may be offered a clerkship in the government patent office in Bern, he writes: "We shall remain students (horribile dictu) as long as we live, and shall not give a damn about the world." Two weeks later, he writes: "Apart from you, all the people look so alien to me as if they were separated from me by an invisible wall."

The young Einstein was especially embittered by his failure to receive recognition from the academic establishment, many of whose eminences he considered self-satisfied men, far beneath him in scientific ability. After sending a letter to Paul Drude, a leading physicist and editor of the prestigious *Annalen der Physik*, about some errors in Drude's work, the twenty-two-year-old Einstein received an unyielding reply from Drude, and wrote to Marić in July 1901:

I have just come home from Lenzburg & found this letter from Drude, which is such an irrefutable evidence of its writer's wretchedness that no comment by me is necessary. From now on I'll not turn any longer to this kind of person but will rather attack them mercilessly via journals, as they deserve. It is no wonder that little by little one becomes a misanthrope.

In November 1901, Einstein submitted a doctoral thesis to Professor Alfred Kleiner at the University of Zurich, criticizing some of the work of the great Ludwig Boltzmann, a colleague of Kleiner's. In December of that year, Einstein wrote to Marić:

> Since that bore Kleiner hasn't answered yet, I am going to drop in on him on Thursday. . . . To think of all the obstacles that these old philistines put in the way of a person who is not of their ilk, it's really ghastly! This sort instinctively considers every intelligent young person as a danger to his frail dignity, this is how it seems to me by now. But if he has the gall to reject my doctoral thesis, then I'll publish his rejection in cold print together with the thesis & he will have made a fool of himself.

ALBERT EINSTEIN was born on March 14, 1879, in Ulm, Germany. His mother, Pauline Koch, was musical and helped get young Albert started in playing the violin. His father, Hermann, was in the electrical business, first with his younger brother Jakob and later on his own, initially investing his own money and then borrowing family money. Hermann suffered one business failure after another. In the hopes of achieving success, the family moved several times, first to Munich in 1880, then to Milan in 1894, to Pavia in 1895, and back to Milan in 1896. Einstein entered primary school in Munich, at the age of six. As his sister, Maja, recalled in her unpublished 1924 biography of her brother, Einstein's teacher taught her students the multiplication tables with the help of whacks on the hand, all to better prepare them early to be good German citizens. Einstein was the only Jew among seventy classmates. One day his teacher of religious studies brought a nail to class and told his students that with such nails Christ had been nailed to the cross by the Jews.

At home, young Albert was becoming more and more of an introvert. Instead of playing with other boys, he liked to work out puzzles, to create complicated constructions out of building blocks, and, with great patience and determination, to build houses of cards many stories high. In the fall of 1888, the boy entered secondary school, the Luitpold Gymnasium. When his family moved to Milan in the summer of 1894, the fifteen-year-old Einstein stayed behind in Munich to finish his studies, boarding with a family. In December, after an argument with one of his teachers at the gymnasium, Einstein abruptly dropped out of school and moved back to Milan with his family. He announced to his parents that he would never return to Munich and would prepare himself to enter college at the Zurich Polytechnic. Then, and for the next few years, he voraciously read physics and mathematics on his own.

After graduation from the polytechnic in 1900, Einstein applied for the position of assistant to most of the leading physics professors in Europe. He was turned down by all of them, possibly because of cool letters from his professor at the polytechnic, Heinrich Friedrich Weber, whom Einstein admired but did not lavish with the customary subservience. (Marić quietly offered her explanation in a letter to a friend: "You know that my sweetheart has a sharp tongue and moreover he's a Jew.") Einstein barely supported himself by tutoring for two years before he received his job at the patent office in Bern, in 1902.

ON JANUARY 6, 1903, Einstein and Marić were married in the stark registry office of Bern—no wedding guests, no members of either family. Serving as witnesses were two friends of the groom, Conrad Habicht and Maurice Solovine, fellow members of the self-appointed "Olympia Academy." For several years, the three friends had been meeting a few nights a week at one of their houses for sausage, cheese, and discussion of Hume,

Kant, Dickens, Poincaré. In November, the newly married couple moved to a rented apartment at 49 Kramgasse, in the old city of Bern. The apartment, on the third floor and reached by a narrow staircase, had only two rooms. However, one of the rooms had large windows looking down on the fine cobblestone street and quaint shops below.

It was here, in the single year of 1905 and working entirely on his own, that the twenty-six-year-old physicist produced four papers that changed the course of physics. Each of the treatises was sufficient to earn him lifetime recognition; one brought him the 1921 Nobel Prize. Two of them presented a mathematical analysis of Brownian motion, the irregular movement of minute particles suspended in a fluid, and provided definitive evidence for the existence of molecules. Another of his papers, which he described to a friend as "very revolutionary," suggested that certain recent experiments could be understood if light flowed in discrete packets of energy, like water droplets, rather than in a continuous stream. The fourth, on special relativity, the greatest achievement of all, proposed a new understanding of time and space.

Only months after these papers appeared in the *Annalen der Physik,* the unknown man at the patent office in Bern was receiving admiring letters from the likes of Phillip Lenard and Max Planck, respectively the leading experimental physicist and the leading theoretical physicist in the German-speaking world. News of the clerk's original thoughts traveled with less velocity in Bern. In April 1906, Einstein's salary at the patent office was increased from 3,900 to 4,500 francs. The physicist attempted but failed to get a better job at the Bern Post and Telegraph Directorate.

Within a few years, however, he received a succession of professorships. In 1909 Einstein was appointed professor at the University of Zurich (where his work on Brownian motion had been accepted for a doctorate four years earlier) and he received his first honorary degree from the University of Geneva. In 1911, Emperor Francis Joseph appointed Einstein a full professor at the German University in Prague; the next year he moved back to a professorship at the Zurich Polytechnic, now called ETH; in 1913 Kaiser Wilhelm II confirmed his appointment to the Prussian Academy of Sciences in Berlin. From then on, Einstein did little formal teaching, which he disliked; in contrast to almost all other academic physicists, great and not so great, he supervised only a single doctoral student in his life. Einstein worked mainly in Berlin until 1933, when he took up his final post as a professor (without duties) at the new Institute for Advanced Study in Princeton.

Since his marriage in 1903, Einstein had been a remote and insensitive husband and father, refusing to take on family responsibilities that might interfere with his work. Months after moving to Berlin, he separated from Marić, who returned to Zurich with their two sons, Hans Albert and Eduard. The divorce settlement in 1919 must be unique in the history of domestic conflicts. In addition to giving Marić custody of the children, Einstein promised her his putative Nobel Prize money, which he and all other physicists on the planet were certain would be awarded. (Such consensus on achievement may be possible only in science.)

The Nobel arrived in 1921. At that time, a journalist gave the following wonderful description of Einstein:

> Einstein is tall (about 1.70 m), with broad shoulders and a scarcely bent back. His head—the head in which the science of the world was newly created—instantly attracts lasting attention. . . . A little mustache, dark and very short, adorns a sensuous mouth, very red, rather big, with its corner betraying a permanent slight smile. But the strongest impression is that of stunning youthfulness, very romantic and at certain moments irresistibly reminiscent of the young Beethoven who, already marked by life, had once been handsome. And then suddenly laughter erupts and one is faced with a student.

EINSTEIN'S NAME was now a household word. In fact, he had rocketed to international fame a few years earlier, in 1919, when observations during a solar eclipse confirmed that light rays from stars were deflected by the gravity of the sun to just the extent he had predicted with his new theory of gravity, general relativity. General relativity was decades ahead of other physicists. It envisioned gravity as a warping of space, the curved orbits of planets as longitudes on a twisting cosmic terrain. Headlines appeared throughout the world. The London *Times* of November 8, 1919, for example, announced: "The Revolution in Science / Einstein Versus Newton." The world was exhausted by World War I, eager for some sign of humankind's nobility, and suddenly here was a scientific genius, seemingly interested only in purely intellectual pursuits. Almost overnight, Einstein became the most honored scientist in history. The new Isaac Newton was invited to address congresses and leaders of state throughout the world; he was given banquets and accolades wherever he went; he was interviewed ceaselessly by the press; great intellectuals and artists, such as Rabindranath Tagore and Sigmund Freud, visited or corresponded with him.

The world was soon to discover that Professor Einstein was not a purely cerebral academic but a vocal intellectual, worthy of comparison to Voltaire as well as to Newton. Beginning in the early 1920s, Einstein drew the hatred of many in his fatherland by his commitments to the political left. He supported the League for Human Rights and, in 1921, helped found the Society of Friends of the New Russia. In 1928, his early pacifism turned militant when he publicly criticized the mostly impotent League of Nations for codifying the rules of warfare: "War simply is no game and cannot be conducted according to the rules of a game. War must be opposed as such, and this can be done most effectively by the masses through an organized wholesale refusal of military service already in peacetime." Two years later, from aboard a ship sailing into American waters, he gave his famous "two percent speech": "If even two percent of those called up declare that they will not serve, and simultaneously demand that all international conflicts be settled in a peaceful manner, governments would be powerless."

Many Germans of the late Weimar period, particularly those preparing to launch a new war, smoldered with anger at these flagrantly unpatriotic statements from the most celebrated German in the world. Yet their hatred could not break into the open, for they needed Einstein more than he needed them. Some of the leading scientists were naturally anxious that he remain in the country. The great physicist Max Planck constantly pleaded with Einstein not to resign his seat in the Prussian Academy of Sciences, and many others in high German society continued to praise him as an exponent of German culture. To all of which Einstein responded with characteristic bite: "A funny lot, these Germans. To them I am a stinking flower, and yet they keep putting me into their buttonhole."

Finally, in 1933, Einstein could no longer allow himself to be stuffed into the German buttonhole and left forever for the United States. He was being denounced in Germany not just for his pacifism but also for his religion. Einstein had never made much of his Jewishness until he went to Berlin and first "understood the Jewish community of destiny," united by a common history of persecution. For Einstein, that common persecution connected with his own life experiences, religious and secular. To a group of Jewish students he later said: "We should be clearly aware of our otherness and draw our conclusions from it. There is no point in trying to convince the others of our emotional and intellectual equality by way of deduction, since the root of their behavior is not located in the cerebrum. Instead we should socially emancipate ourselves and essentially satisfy our social requirements ourselves."

In 1935, Einstein and his second wife, Elsa Löwenthal, moved to an inconspicuous house at 112 Mercer Street in Princeton, a New England clapboard with a small garden in front and a long narrow garden in back. Einstein took as his study a room on the second floor, with a large picture window that looked across a wooded park to the university's Gothic buildings in the distance. It was here that Einstein spent most of his time, and was the happiest, in the last two decades of his life.

TWO

AT THE TURN of the century, when the young Einstein began crafting the theory of special relativity, physicists knew of two fundamental forces of nature: the gravitational force and the electromagnetic force. (Electrical and magnetic forces were considered parts of the same force because they could generate each other, as a wire twirling near a magnet generates an electrical current.) Einstein did not initially set out to propose new concepts for time and space but rather to explain a particular problem in the theory of electromagnetism.

The problem was this: during the last decades of the nineteenth century, physicists had shown that light consisted of undulating waves of electrical and magnetic energy traveling through space. All known undulations of energy, such as sound waves, required a material substance to traverse. For example, when sound travels across a room, it is the air molecules in the room, bumping successively into one another like a row of falling dominoes, that constitute the traveling "wave" of sound. Sound can travel through a gas, a liquid, or a solid, but it cannot travel through a vacuum. It needs the dominoes. By analogy, scientists assumed that the propagation of light required an underlying material substance. This substance was called the ether. Furthermore, the ether, a gossamer and practically weightless substratum, was required to fill up every nook of the universe, since the received glimmer of distant stars clearly showed that light could travel through the vast reaches of space as well as across a living room.

Now, it is both intuitively obvious and also true that sound will travel more quickly across a room if it rides not on static air but on a wind blowing through the room. By analogy, physicists assumed that the speed of light should depend on the motion of the ether going past the observer or the observer going past the ether. (In physics, an "observer" is anyone equipped to make measurements.) The great quandary was that all attempts to measure the motion of an observer through the ether had proved unsuccessful. Light rays seemed to travel at the same speed regardless of which way the assumed ether "wind" was blowing. Electrical and magnetic experiments performed in a train station gave results identical to the same experiments performed in a train traveling by the station at sixty miles per hour.

The ether's physical reality had never been confirmed. Einstein postulated that it did not exist, that it was a superfluous construction. Without an ether, it could not be said that an observer was moving or not with respect to the ether. In fact, it could not be said that an individual observer was moving at all, in absolute terms. By postulate, Einstein had eliminated the cosmic frame of reference against which all individual motions could be measured. Only the relative motion of two observers, or any two things, had meaning. Hence the name "relativity." Correspondingly, all observers could be considered as equivalent. All observers, as long as they were not accelerating, would produce identical measurements for all electrical and magnetic phenomena. In particular, a man sitting on a couch would measure exactly the same speed for a passing light ray as a man running by him in the direction of the ray.

To permit himself this last seemingly preposterous proposition, Einstein first recognized that our notions of time and distance, both of which enter into the measurement of speed, had not been carefully analyzed before. Anchored in the unconscious, for example, was a belief in the absolute character of time. A second for me is a second for you. As a practical matter, the possible discrepancies between clocks moving past each other at the slow speeds of earthly life would be much too small to have been noticed. Philosophically, Einstein disagreed with Kant, whom he had read as a youth. He opposed Kant's view that particular concepts of time and space are necessary premises for thought, innate in the human mind. For Einstein, all concepts, even physical ones, are "free inventions of the human intellect," which can be modified if they do not prove useful in comparison to experience. The meaning of space and time (for physics) is determined, he wrote, by considering the spatial separation and time elapsed between two events, like two successive chimes of a grandfather clock. When observers in relative motion measure these same two events, their measuring sticks and clocks must be calibrated and synchronized in a reproducible way.

Starting with these working definitions of time and space, and the postulate that there was no ether, Einstein calculated quantitatively how the ticking rates of clocks and the lengths of measuring sticks in motion with respect to each other must differ so that both sets of instruments measure the same speed for a passing ray of light. The predicted differences are indeed tiny for everyday relative motions—one second on a supersonic jet takes 1.0000000000014 seconds as measured on the ground—but the differences have indeed been detected and agree with the theory. For relative speeds closer to the speed of light, 186,000 miles per second, such as occur in high-energy particle accelerators, the effects are much larger. A second for me is not a second for you. However, the deep significance of special relativity extends far beyond its quantitative results. More than any philosopher or scientist before him, Einstein demonstrated that our intuition about the physical world based on sensory experience can be fundamentally in error.

HOW MUCH DID Einstein's temperament, especially his sense of being an outsider, affect his science? Here, of course, one can only point to certain suggestive associations rather than prove causal relationships.

First of all, Einstein used a rare, deductive approach to science. In this he differed from most other scientists of the time, including other theoretical physicists. In the more usual inductive approach, the scientist begins with a number of observations about nature, tries to find a pattern, generalizes the pattern into a "law" or organizing principle, and then tests this law against future experiments. For example, the German astronomer Johannes Kepler examined the data on planets, analyzing the numbers in many different ways, before discovering a striking relationship between a planet's distance from the sun and the time it takes to complete an orbit. This brief account of the methodological process is certainly oversimplified, and the work is often carried out by an entire community of scientists rather than by a single person. Essentially, however, experiments, data, and observations form both the starting point and the center in inductive science. In deductive science, on the other hand, the scientist begins by postulating certain things to be true, with only secondary guidance from outside experiments, and then deduces the consequences of those postulates. Finally, the consequences are tested experimentally. If the tests fail, the postulates must be changed.

In his *Autobiographical Notes* (1946), Einstein explicitly credits the philosopher David Hume for having taught him that the truths of nature cannot be arrived at by experience with the world. Rather, one must start with the "free inventions" of the mind. In the case of special relativity, for example, Einstein began with the postulate that the ether did not exist, that absolute motion did not exist, that all unaccelerated observers would measure identical laws of physics; he then derived the consequences of those ideas. Other theoretical physicists of the time, most notably the Dutch Hendrik Antoon Lorentz, proposed detailed theories for how bodies moving through the ether would be electrically compressed in exactly the right amount, depending on their speed, in order to make the experimental results come out as they did. It seems possible that the depth of

Einstein's inner world made him especially attentive to Hume's teachings and naturally inclined to explore nature from within his own mind.

TWO ASPECTS of Einstein's later scientific career are suggestive of his stubborn self-confidence and willingness to strike out completely on his own: his philosophical refusal to accept the new quantum physics, and his continued work on his own unified theory (combining gravity and electromagnetism), despite one failure after another.

A fundamental tenet of quantum theory, developed in the 1920s, is that the position and velocity of an individual particle cannot be completely specified, even in principle. As a result, one cannot predict with certainty the future position and velocity of a particle; such predictions can be done only in terms of probabilities, which apply to the average behavior of a large number of particles. Einstein stridently opposed this indeterminacy inherent in quantum theory. In his view, everything, even the behavior of a single electron, should be deterministic and calculable. In a letter to his fellow physicist Max Born, Einstein wrote: "The idea that an electron exposed to a ray by *its own free decision* chooses the moment and the direction in which it wants to eject is intolerable to me. If that is so, I'd rather be a cobbler or a clerk in a gambling casino than a physicist."

While the current of physics swerved in the direction of the powerful new quantum physics, which was able to explain the behavior of atoms, Einstein became obsessed with his nonquantum unified theory, which he doggedly pursued in increasing isolation for the rest of his life. With each new attempt to perfect that theory, the physicist was sure that he was closer to victory. ("The latest results are so beautiful that I have every confidence . . .") The discovery of antimatter in the 1930s and of new subatomic particles, hinting at new fundamental forces, all passed by with little comment from Einstein. Even when the great Danish physicist Niels Bohr, with whom he had once had lively discussions, visited the Institute for Advanced Study in 1939, Einstein remained cloistered in his paper-strewn room at the institute.

Einstein's brittleness here cannot be attributed simply to age. Born and Bohr, for example, were only a few years younger than Einstein, yet both immediately embraced the new quantum theory and became leading practitioners of it. In a letter to his friend and former physician Otto Juliusburger in 1937, Einstein wrote: "I am not really becoming part of the human world here [in Princeton], for that I was too old when I arrived, and in point of fact it was no different in Berlin or in Switzer-

land. One is born a loner." Einstein's independence and stubbornness, which had served him so well in the first part of his scientific career, became a curse in the last.

WHEN ONE LOOKS at Einstein as a human being, one sees a man of contradictions. Einstein scoffed at material concerns, yet he consistently demanded exorbitant salaries and lecture fees once he became famous. He opposed nationalism of all kinds, including Jewish nationalism, but he supported the birth of Israel and later the founding of the Hebrew University. He was an adamant pacifist who scorned the German military mentality. Yet he remained friends with Fritz Haber, who inaugurated chemical warfare and the use of poison gas in 1915 (the year Einstein published his paper on general relativity), and in fact for some time maintained his office in Haber's Kaiser Wilhelm Institute. To his friend H. Zangger, Einstein wrote in 1917: "Our entire muchpraised technological progress, and civilization generally, could be compared to an axe in the hand of a pathological criminal." Yet during this same period Einstein worked on the improvement of aircraft wings and collaborated with the rich young inventor Hermann Anschütz-Kaempfe on the first gyrocompass, used on German U-boats during the war. (By the 1930s, almost every navy in the world, except the British and American, steered their courses with gyrocompasses based in part on a patent owned by Einstein.)

Einstein the humanitarian lent his name to social causes and, once the great wave of European Jewish emigration began in the mid-1930s, he personally supported and financed the families of German friends in their relocations in America. After the German annexation of Austria in 1938, he tried to start a relief organization supported by churches, universities, and the Red Cross. Yet in other circumstances the same man seemed an elitist. To his cousin Elsa, whom he was to marry six years later, he wrote in 1913: "Traveling people, both of us, destined to high-wire dancing from among the swarm of Philistines . . . ," and "I have firmly resolved to . . . walk only in really agreeable company, in other words rarely." In 1931, he wrote of the death penalty: "I would have no objection to the killing of worthless or even harmful individuals; I am against it only because I do not trust people, i.e., the courts. I appreciate more the quality than the quantity of human life." In the publication *Forum and Century* the same year: "The really valuable thing in the pageant of human life seems to me . . . the creative, sentient individual . . . while the herd as such remains dull in thought and dull in feeling."

Einstein was both a man of high principles and an opportunist, a loner and an activist, a liberal and an elitist, a great theoretician and a practical examiner of patents. By the time he was forty, he was well aware of his high place in history. Yet he showed himself capable of humility and kindness. "A hundred times every day I remind myself that my inner and outer life are based on the labors of other men, living and dead," he once wrote. In Einstein's last decade of life, his ophthalmologist in Princeton checked his eyes annually and invariably told the old man that his glasses would be much improved with a new prescription. And Einstein would invariably reply with a smile: "A friend in New York sends me these simple magnifying glasses as a gift each year, and if they do no real harm, Henry, I prefer not to change them. I don't want to hurt his feelings."

(1997)

THE ONE AND ONLY

RICHARD FEYNMAN was the Michael Jordan of phys-ics. His intellectual leaps, seemingly weightless, defied explanation. One of his teammates on the high-school math team in Far Rockaway, Long Island, recalls that Feynman "would get this unstudied insight while the problem was still being read out, and his opponents, before they could begin to compute, would see him ostentatiously write down a single number and draw a circle around it. Then he would let out a loud sigh." At twenty-three, he amazed a Princeton colleague when he worked out at the blackboard a proof of an important proposition of physics that had been only loosely con-jectured eight years earlier by the Nobel Prize winner Paul Dirac. In 1960, in his early forties, restless and unable to find a physics problem worth working on, Feynman taught himself enough biology to make an original discovery of how mutations work in genes.

Feynman rarely read the scientific literature. When he did, he would read only far enough into an article to see what the problem was, fold up the journal, and then derive the results on his own. When a colleague, after perhaps months of calculations, walked into Feynman's office with a new result, he would often discover that Feynman already knew not only that result, but a more sweeping one, which he had kept in his file drawer and regarded as not worth publishing. The mathematician Mark Kac has said that "there are two kinds of geniuses, the ordinary and the magicians. An ordinary genius is a fellow that you and I would be just as good as, if we were only many times better." But for the second kind, "even after we understand what they have done, the process by which they have done it is completely dark. . . ." He called Feynman "a magician of the highest caliber."

SCIENTIFIC GENIUS ALONE would not have explained Feynman's legend. It was also his style. He was stubborn, irreverent, unrefined, uncultured, proud, playful, intensely curious, and highly original in practically everything he did. He had a mystique. There are hundreds of "Feynman stories," some told by Feynman himself in his popular book *"Surely You're Joking, Mr. Feynman!"*; many documented in James Gleick's superb biography of Feynman, *Genius;* and others passed along by word of mouth from one physicist to another, like beheld visitations passed from one disciple to another.

As a graduate student at Princeton, for example, Feynman would spend long afternoons leading ants to a box of sugar suspended by a string, in an attempt to learn how ants communicate. When Feynman noticed that his Ph.D. thesis adviser, John Wheeler, pointedly placed his pocket watch down on a table during their first meeting, Feynman came to their second meeting with a cheap pocket watch of his own and placed it on the table next to Wheeler's. At Los Alamos, when he was working on the Manhattan Project, the young Feynman continually alarmed other scientists and the military brass by cracking their safes, which were filled with atomic secrets.

When he was preparing to accept the Nobel Prize in the presence of the king of Sweden, Feynman worried that it was forbidden to turn one's back on a king; he might, he was told, have to back up a flight of stairs. He then practiced jumping up steps backward, using both feet at once. Feynman hated pomp and authority of all kinds. After being elected to the prestigious and highly selective National Academy of Sciences, he withdrew from the organization, saying that its main function was only to elevate people to its exalted ranks.

There was something almost uncanny about the way Feynman could get to the heart of a question. On February 10, 1986, during the public hearings on the *Challenger* shuttle disaster, as a member of the committee of inquiry, he performed an experiment of deadly simplicity. He dropped one of the shuttle's O-ring seals into a glass of ice water, the temperature of the air on the day of the launch, and showed that the rubber when squeezed did not stretch back under such cold.

Feynman was born in New York on May 11, 1918. His father, Melville, a Jewish immigrant from Minsk, Byelorussia, had a practical, vivid appreciation of science; he once explained to his son Richard that a dinosaur twenty-five feet high with a head six feet across, if standing in the front yard, would almost be able to get his head through the second-floor window. Melville Feynman sold police uniforms and automobile polish, among other ventures, without notable success. Feynman's mother, Lucille, had a gift for humor and a love of storytelling.

As a child in Far Rockaway, Feynman tinkered with radio sets, gathering spare parts from around the neighborhood. Many theoretical physicists like Feynman have spent their childhoods building things, but Feynman retained throughout his life an immediate, tactile sense of physical phenomena. Even his mathematical calculations have a certain unfussy and muscular style. Feynman rigged a motor to rock his sister's crib, freeing himself to read the *Encyclopaedia Britannica*. But he was intimidated by athletics, by stronger boys, and by girls, and was afraid that he would be regarded as an intellectual sissy. Like so many other socially fragile, budding scientists he sought refuge in an intense concentration on math and science, but he was particularly interested in their practical side. His manliness, he saw, lay in his ability to do things with his hands.

Correspondingly, he avoided all pursuits that seemed to him "delicate," such as poetry, drawing, literature, and music. In fact, Feynman had little respect for the humanities, which he regarded as slippery and inferior to science, and even less respect for humanists. When he was in his early thirties, he wrote that "the theoretical broadening which comes from having many humanities subjects on the campus is offset by the general dopiness of the people who study these things." Yet Feynman had an appreciation of the workings of human psychology in science. In his brilliant little book *The Character of Physical Law* he places great value on seeking different formulations of the same physical law, even if they are exactly equivalent mathematically, because different versions bring to mind different mental pictures and thus help in making discoveries. "Psychologically they are different because they are completely unequivalent when you are trying to guess new laws."

IN THE FALL OF 1935, Feynman entered MIT, where he found that virtually everyone else was socially and athletically inept, and obsessed by science. He easily skipped first-year calculus and taught himself quantum mechanics before his sophomore year. He joined a fraternity, one of the two that took in Jews. He met another precocious physics student, T. A. Welton, and together they rederived for their own satisfaction the basic results of quantum physics, which they wrote down in a notebook that they passed back and forth to each other. Feynman briefly read Descartes and decided that philosophy was soft and that philosophers were incompetent logicians. It was in his junior year at MIT that he became engaged to Arline Greenbaum, whom he had met a few years earlier in Far Rockaway, and who became, besides physics, the love of his life.

Where MIT was working-class in tone and unbuttoned in manner, Princeton was patrician and genteel. The afternoon Feynman arrived there as a graduate student, in the fall of 1939, he was invited to a tea with Dean Eisenhart. As he stood uneasily in the suit he hardly ever wore, the dean's wife, a lioness of Princeton society, said to him, "Would you like cream or lemon in your tea, sir?" "Both please," Feynman blurted out. "Surely you're joking, Mr. Feynman," said Mrs. Eisenhart, thus supplying the title for the memoir Feynman published fifty years later. Feynman hated people who, he felt, used manners and culture to make him feel small. He became aggressively unrefined.

Feynman and his thesis adviser at Princeton, John Wheeler, worked on the nature of time. Einstein had already shown, at the beginning of the century, that time is not absolute, that the rate at which clocks tick depends on the motion of the observer. But what determines the direction of time? Why is the future so distinct from the past? It was well known that, at the microscopic level, the laws of physics were indifferent to the direction of time: they gave the same results whether time flowed forward or backward. Feynman and Wheeler solved a difficult problem concerning electricity by assuming that, in the way electrons emit radiation, time flows both forward and backward. It seemed a crazy idea, but it was the kind of deep and important crazy idea that caused physicists to skip meals and stay at the blackboard. By the time the young Feynman presented his calculations to a departmental seminar in early 1941, his audience had come to include the great mathematician John von Neumann, the physicist Wolfgang Pauli, who was soon to win the Nobel Prize and was visiting from Zurich, and the sixty-two-year-old Einstein, who seldom came to colloquia. After listening to Feynman's talk, Einstein commented, in his soft voice, that the theory seemed possible.

AROUND THIS TIME, Arline, who had been suffering from fevers and fatigue, was diagnosed as having tuberculosis. She was to spend much of the rest of her short life in sanatoriums. Against his parents' strong objections, Feynman married her. The only witnesses to the wedding, in a city office on Staten Island, were two strangers.

In 1942, many of the physicists at Princeton began fanning out to work on military projects at, among other places, MIT's Radiation Laboratory (the "Rad Lab"), the University of Chicago, Berkeley, and Oak Ridge, Tennessee. While still at Princeton, Feynman collaborated with Paul Olum and Robert Wilson on a device for culling the fissionable form of uranium from the nonfissionable. This was the beginning of the Manhattan Project. In March 1943, Feynman and Arline took the train to Los Alamos. Arline entered Presbyterian Sanatorium in Sante Fe while Feynman lived in the barracks at Los Alamos, driving the twenty-five miles over rutted roads to see her every weekend.

At Los Alamos, Hans Bethe, the great nuclear physicist from Cornell, was in charge of all theoretical work. Where Bethe was calm, careful, and professorial, Feynman was quick, fearless, intuitive, and irreverent. Feynman was just what Bethe was looking for. He made the twenty-five-year-old Feynman a group leader, promoting him over older and more senior physicists.

Feynman was able to solve a critical problem on how neutrons bounce around among uranium atoms and start a chain reaction.

Arline died in the summer of 1945. Two years later, when he was at Cornell during a frustrating impasse in his theoretical work, the twenty-nine-year-old Feynman wrote a letter to his dead wife, placed it in a box, and never read it again. After Feynman's death, his biographer Gleick discovered the letter, which reads in part:

D'Arline,

I adore you, sweetheart.

It is such a terribly long time since I last wrote to you—almost two years but I know you'll excuse me because you understand how I am, stubborn and realistic; & I thought there was no sense to writing.

But now I know my darling wife that it is right to do what I have delayed in doing. . . . I want to tell you I love you. I want to love you. I always will love you.

I find it hard to understand in my mind what it means to love you after you are dead—but I still want to comfort and take care of you—and I want you to love me and care for me. I want to have problems to discuss with you. . . .

P.S. Please excuse my not mailing this—but I don't know your new address.

Arline's death was the great tragedy of Feynman's life. Feynman seems to have never let anyone get close to him again, although he had many affairs, a second brief and unpleasant marriage, and a third, apparently satisfying one to Gweneth Howarth, with whom he had two children. For the rest of his life, Feynman pursued only beautiful women, some of them the wives of his friends and colleagues, but he had no interest in their intellectual companionship. His attitude toward women is suggested by the conclusion of his Nobel address in 1965:

> So what happened to the old theory that I fell in love with as a youth? Well, I would say it's become an old lady, that has very little attractive left in her and the young today will not have their hearts pound when they look at her anymore. But, we can say the best we can for any old woman, that she has been a good mother and she has given birth to some very good children.

Beginning in his late twenties, Feynman started to be followed around by Feynman stories. He heard the stories, polished and embellished them, and retold them. He relished his image as a rough-hewn, philistine hero. Gleick writes that after Arline's death, "The Feynman who could be wracked by strong emotion, the man stung by shyness, insecurity, anger, worry or grief—no one got close enough any more to see him."

Feynman eventually emerged from his depression at Cornell and, in the late 1940s, did the work that won him the Nobel Prize, showing how electrons interact with electromagnetic radiation—e.g., radio waves—and other charged particles. His theory, called quantum electrodynamics, has been confirmed by experiments to greater accuracy than any other theory of nature. (Quantum electrodynamics predicts that the magnetic strength of the electron is 1.00115965246; the measured value is 1.00115965221.) Quantum electrodynamics explains *all* electrical and magnetic phenomena, which include everything we experience in daily life except gravity.

Feynman shared his prize with Shin'ichirō Tomonaga of Japan and with Julian Schwinger of the United States, who had both independently derived their own formulations of quantum electrodynamics. These alternative formulations were, however, much harder to work with than Feynman's. Schwinger was in many ways the antiparticle of Feynman. He dressed expensively and meticulously, drove a black Cadillac, spoke elegantly in long sentences with subordinate clauses, lectured without notes, and prided himself on arriving at the end of complex mathematical calculations with no dust on his shoes from taking an occasional blind alley.

FEYNMAN MADE two other major contributions to physics, both worthy of a Nobel Prize. He developed a theoretical explanation for superfluids—fluids that are totally frictionless and that will spontaneously glide over the walls of a beaker and will pass through holes so tiny that even gas could not get through. He also worked out a theory for the weak force, one of the four kinds of fundamental forces. Both theories were developed at the California Institute of Technology, where he spent the second half of his life. Murray Gell-Mann, Feynman's rival at Caltech, had independently arrived at the weak-force theory, and the department chairman judiciously arranged for Feynman and Gell-Mann to publish their important work in a joint paper. Like Schwinger, Gell-Mann was very different from Feynman. His interests in science were narrow, but he had broad interests outside science, while Feynman was engrossed with virtually all of science, but with almost nothing outside it.

At Caltech, Feynman became more concerned with education, although he did not have the patience to supervise students preparing theses. In 1961, Caltech decided to revise its physics curriculum and asked Feynman to help. Lecturing at the blackboard to freshmen and later to sophomores, he began with atoms, moving up to larger phenomena like clouds and colors on ponds and down to the smaller, like electrons and the quantum world. Without consulting books, he slowly built up the entire edifice of physics as he understood it, the physical world as he saw it. Soon graduate students and other professors came to listen. Feynman's Caltech lectures eventually became the three-volume *Feynman Lectures on Physics,* which can be found on the bookshelves of almost every professional physicist in the world. The lectures ultimately failed to accomplish their intended purpose. Apparently simple on the surface, they were in fact deeply sophisticated. But they are a triumph of human thought and deserve a place in the history of Western culture, along with Aristotle's collected works, Descartes's *Principles of Philosophy,* and Newton's *Principia.*

FEYNMAN WON the Nobel Prize for his work in quantum electrodynamics, the quantum theory of how electrons interact with radiation and other electrically charged particles. Electrons are the simplest electrical particles. Normally found in the outer parts of atoms, they produce light and other forms of electromagnetic radiation as well as most of the interactions between atoms and molecules. Quantum physics, one of the two pillars of twentieth-century physics along with Einstein's relativity, is the physics of the subatomic world. The theoretical foundations of quantum physics were laid in the 1920s, principally by Erwin Schrödinger, Werner Heisenberg, and Paul Dirac. A basic idea of quantum physics is that particles of matter sometimes behave as if they were in several places at once. This uncertainty about the location of things is negligible for macroscopic objects like people, but it is extremely important for subatomic particles like electrons, where the phenomenon has been repeatedly observed and has immense consequences. Another important idea, also derived from experiment, is that physical quantities like energy are not indefinitely divisible into smaller amounts, but instead have a smallest, indivisible unit, called the quantum (as U.S. currency has a smallest unit, the penny). Both ideas run counter not only to intuition but to the Newtonian theoretical conception of the world before 1900. In order to mathematically describe these two basic ideas of quantum physics, the theories of Schrödinger, Heisenberg, and Dirac had to represent matter and energy not by certainties but by probabilities (or, technically speaking, amplitudes, which are closely related to probabilities). Thus, while in the Newtonian scheme a physical law would show how a particle moves from A to B under the action of a force, in the quantum scheme a physical law would show how the probabilities for a particle to be at various places evolve under the action of a force.

The quantum theory of the 1920s gave a good description of isolated particles, but it did not accurately describe the interactions of particles. Experiments began to turn up small discrepancies in particle behavior. For example, in the strange quantum world, subatomic particles are constantly appearing out of nothing and then disappearing again. Each particle, such as the electron, surrounds itself with a cloud of other, ghostlike particles, called "virtual particles," which fleetingly come into existence and then slip away into oblivion. Electrons interact with the ghostlike particles around them, and those interactions alter the properties of the electron, such as its mass and electrical charge. In reality, the physicists found, there are no isolated electrons. The quantum ghosts are everywhere. Their shadows have been seen in experiments. When physicists in the late 1930s and early 1940s tried to modify the quantum theory of Schrödinger, Heisenberg, and Dirac so as to accurately describe particle interactions, they ran into technical difficulties with the ghosts. Once the ghosts began popping up in the mathematics, the equations couldn't be solved.

One of the triumphs of Feynman's quantum electrodynamics was that it provided a method for dealing with the ghosts. Roughly speaking, the method involves treating the ghosts as part of the electron. Experiments on electrons do not penetrate inside the cloud of ghost particles around them; we never observe the "bare" electron at the center of the cloud. What we observe is the electron *and* its cloud. When the thing we call an electron is redefined to include the virtual particles around it, the technical difficulties go away.

Other scientists in addition to Feynman contributed to this redefinition of the electron (and its subatomic cousins). However, Feynman's own version of quantum electrodynamics had two further, and unique, features. First, it made use of mathematical methods that were much easier to work with than the methods of other versions, particularly the version of Schwinger. This was Feynman at his practical best. Second, Feynman's quantum electrodynamics provided a new picture of the world. In other descriptions of quantum physics, even after certainties are replaced by probabilities, a particle advances from A to B in tiny increments, with forces acting to move the particle (or the probability of the particle) from one increment to the next. But Feynman's mathematical description of quantum electrodynamics is global, not incremental. It considers every possible route from A to B, assigns a single number to the *entire* route, then adds up the numbers from all the different routes to arrive at the probability of getting from A to B.

The descriptions of other physicists could be compared to observing how a car speeds up and slows every few feet along a highway from New York to Los Angeles, whereas Feynman's description looked only at the total gas consumption for the trip. Furthermore, in Feynman's description, the car travels simultaneously on all routes from New York to Los Angeles, even on such crazy but possible routes as New York to Chicago to Miami to Los Angeles. Such a description leads to a strange picture of the world, where all the different ways in which something can happen *are* happening, at the same time. What we human beings—grossly insensitive, macroscopic objects that we are—conceive of as a single reality is actually a tapestry of many simultaneous realities. It is ironic that Feynman, who considered philosophy a waste of time, should have come up with ideas philosophically so rich. But all deep theories of nature since Lucretius's atomism have had broad philosophical implications.

IN FEBRUARY 1988, after a gruesome series of illnesses and complications from cancer, Feynman entered the UCLA Medical Center for the last time. He was sixty-nine. Across the city, on a corner of his blackboard, he had written in chalk, "What I cannot create I do not understand." As he lay in his hospital bed, with his strength ebbing, Feynman whispered his last words: "I'd hate to die twice. It's so boring."

(1992)

MEGATON MAN

ONE

IN LATE 1951, on an overnight train from Chicago to Washington, Edward Teller dreamed that he was alone, in a battlefield trench like the ones that had so terrified him as a child in Hungary during the war. The nine men attacking his position exceeded by one the eight bullets in his rifle—a cold mathematical analysis even in the confused and foggy world of a nightmare.

Teller's dream might be simply related to anxiety over his impending report to a subcommittee of the Atomic Energy Commission (AEC), where he was lobbying for the creation of a new weapons laboratory. Yet more deeply the dream expresses a lifelong sense of being embattled, besieged, alone in a righteous struggle against his many enemies and the forces of evil. Teller remembers being insulted by his ninth-grade mathematics teacher when he correctly answered a question based on material not yet covered in class. "What are you? A repeater?" said the teacher. The boy prodigy was never called on again, even when he was the only one to raise his hand. While working at Los Alamos on the Manhattan Project, where he pursued his own projects rather than his team's assignments, Teller "slowly came to realize . . . that my views differed from those held by the majority" in his fear of Communist Russia and in his fierce support of an overwhelming American military superiority extending far beyond World War II.

Soon Teller's friendship with Robert Oppenheimer and Hans Bethe, both eminent colleagues at Los Alamos, soured as they engaged in mutual criticism, a pattern that was to repeat itself throughout Teller's life. After the successful construction of the atomic bomb and the end of the war, when Oppenheimer, Bethe, and many other physicists returned to university teaching and peacetime work, Teller felt that he was a lone voice in pushing the development of the hydrogen bomb; leading scientists, he believed, were "trying to prove a hydrogen bomb impossible." He much resented Norris Bradbury, the new director of the Los Alamos weapons laboratory (replacing Oppenheimer), for dragging his feet on the hydrogen weapon, called "the Super" because of its potentially unlimited power and destructiveness; he claimed that Carson Mark, the new head of the theory division (the position Bethe had held), "made it a practice to needle me in a subtle manner." Everywhere Teller turned, it seemed, were enemies and suspicions.

Teller's fragile link to his colleagues was finally broken by his hugely unpopular testimony against Robert Oppenheimer in the McCarthy-era hearings of 1954, which deprived the brilliant and charismatic Oppenheimer of his security clearance and forever excommunicated Teller from most of the scientific community. Shortly after the hearings, when Teller spotted a long-time physicist friend at a meeting and hurried over to greet him, "he looked me coldly in the eye, refused my hand, and turned away." Twice before, oppressive governments and anti-Semitism had driven Teller into exile, from Hungary in early 1926 and from Germany in 1933. "Now, at forty-seven," he recalls, "I was again forced into exile." Years later, after countless political intrigues, after battling with scientists and politicians alike for his proposed projects ranging from nuclear energy to nuclear explosives for excavations to an antimissile defense system, Teller writes that he finally learned a slogan for life: "Trust nobody."

OF THE GREAT physicists who ushered in the modern age of the atom, only three remain: Edward Teller, age ninety-four, Hans Bethe, age ninety-five, and John

Wheeler, age ninety. Gone are Ernest Rutherford, James Chadwick, Niels Bohr, Werner Heisenberg, Lise Meitner, Otto Hahn, Eugene Wigner, Enrico Fermi, and many others. Compared to Teller, the meticulous Bethe and the self-effacing Wheeler have lived quiet lives in a monastery. In his towering public persona and impact, Teller is equaled by only a handful of twentieth-century scientists: Albert Einstein, Linus Pauling, and James Watson among them. In his siege mentality and violent controversies, Teller stands alone. One of the creators of the new quantum physics, a principal architect of the hydrogen bomb, founder and guiding force of the giant Livermore weapons laboratory, passionate advocate of nuclear power and antimissile defense, hypnotic teacher and lecturer, amateur pianist and performer of Beethoven and Bach, student of Plato—Edward Teller, whatever one's attitude toward his politics, his bullying tactics and prevarications, must be regarded as a man of vision and staggering accomplishments.

An incident in 1962, which Teller proudly relates, illustrates his power. The occasion was his invitation to the Southern Governors' Conference to argue against the pending Limited Test Ban Treaty. (Throughout his career, Teller staunchly opposed all nuclear weapons treaties.) On his arrival in Arkansas, barely awake from yet another night on a train, the Hungarian physicist was informed that President Kennedy had sent a worried message to the governors protesting Teller's presentation on the grounds that there was no one at the conference to rebut him. Elsewhere, Teller credits the huge weapons stockpile he helped to create with preventing World War III. Now, crippled with arthritis and suffering from macular degeneration, Teller writes: "I am not about to stop working; I still have many projects to complete and an infinite number of problems to address."

His new autobiography, *Memoirs,* far more comprehensive than his 1962 *The Legacy of Hiroshima,* testifies to his astonishing stamina and mental facility, as well as sharp wit. Some of the recollections in *Memoirs* are directly contradicted by the accounts of other people; some are merely embroidered or skewed. Fortunately, a host of critical accounts, such as Richard Rhodes's *The Making of the Atomic Bomb* and *Dark Sun* and William Broad's *Teller's War,* and a great many documents allow us to make some judgment about the real Edward Teller. What I have come away with, after sifting through the numerous inconsistencies and contradictions, is that there are two Edward Tellers. There is a warm, vulnerable, honestly conflicted, idealistic Teller, and there is a maniacal, dangerous, and devious Teller. Moreover, like Dr. Jekyll, Teller is disturbingly aware of his darker side. Indeed, that self-awareness, visible in *Memoirs* even beneath its fabrications and self-congratulation, is what accounts for Edward Teller's angst and gives him his true tragic proportions.

TWO

EDWARD TELLER was born on January 15, 1908, in Budapest. His father was a lawyer and associate editor of the major law journal of Hungary. In a footnote, Teller comments that his father, a quiet and reserved man, "did all of the routine work" on the journal while the chief editor "added the flair." One cannot refrain from speculating whether Teller's own later style, avoiding routine work and spraying out gallons of original ideas, might have been some unconscious reaction against the tedium of his father's life.

Teller recalls that "finding the consistency of numbers is the first memory I have of feeling secure." That security was challenged again and again. When the Communists briefly took over Hungary in 1919, Teller's father was considered a capitalist and the family became social outcasts. This ordeal was the eleven-year-old Teller's first taste of communism. Years later, in 1939, Teller's hatred of communism became fierce when his friend the Russian physicist Lev Landau was sent to prison by Stalin for imagined disloyalty and emerged a year later a broken man. And in 1962, at the height of the cold war, Teller would write: "In Russian Communism we have met an opponent that is more powerful, more patient, and incomparably more dangerous than German Nazism."

Teller detested his years at high school, where his classmates laughed at him and nicknamed him Coco, meaning a simpleminded clown. When Teller wanted to transfer to another school, he was turned down because he was not Catholic. "I began to wonder whether being a Jew really was synonymous with being an undesirably different kind of person."

The young Teller became possessed by science. In the fall of 1929, he moved to Leipzig to begin his doctoral work under Werner Heisenberg, one of the founders of quantum mechanics and already a legend at age twenty-seven. Here, Teller joined an international group of twenty eager young men. Once a week, Heisenberg's disciples met for an evening of Ping-Pong, chess, and tea. Seven days a week, they argued about physics, art, and life.

Teller and Heisenberg formed a close bond. The mentor and his apprentice took turns playing preludes and fugues from Bach's *Well-Tempered Clavichord* on the excellent grand piano in Heisenberg's apartment. After World War II, when most of the scientific community harshly condemned Heisenberg for his attempt to build an atomic bomb for the Nazis, Teller alone claimed that

Heisenberg was innocent, writing generously and perhaps naively that "it is inconceivable to me that Heisenberg would ever have pursued such a [weapon]. He loved his country, but he hated the Nazis."

During the next few years Teller married his childhood sweetheart, spent a year in Copenhagen with the great Niels Bohr, and worked for a couple of years at University College London. In his memories of these early scientific associations, Teller shows himself to be a keen observer of people, including himself. Teller, and indeed all physicists, revered Bohr, the father of the first quantum model of the atom, a gentle man who spoke so softly that he could scarcely be heard:

> Bohr invented paradoxes because he loved them. I imagine that I understand those paradoxes, but I failed to understand Bohr. In human terms, understanding means being able to put yourself in the place of a fellow being. In those terms, I can understand Heisenberg; if my abilities were much greater than they are, I could imagine myself in his position. In no way can I imagine myself in Bohr's place.

Two recollections of this youthful period in Copenhagen reveal Teller's recognition of his own hotheaded nature and his tendency to inflate the truth, problems that would cause him and others grave difficulties throughout his life. When Teller once pointed out a silly overstatement that Bohr had made in a casual remark, the Danish physicist replied: "If I can't exaggerate, I can't talk." The ninety-four-year-old memoirist Teller writes: "I have quoted [Bohr] in many discussions in defending my own right to exaggerate." But Teller's exaggerations would be of more consequence, linked as they were to the nuclear arms race and costing billions of dollars. Teller recalls discussing Aristotle's classification of different types of personality with Carl Friedrich von Weizsäcker, a German physicist who was later to work with Heisenberg on the German A-bomb. "Carl Friedrich," he writes, "correctly named me, not as sanguine as some of my critics have claimed, nor as melancholic, as I sometimes feel, but as choleric, a flaw I struggled against in my youth."

AT THE INVITATION of the physicist George Gamow, Teller came, in 1935, to the United States, where he would spend the rest of his life, first at George Washington University for five years, then at the University of Chicago, the Los Alamos Laboratory, the Livermore Laboratory, Berkeley, and finally the Hoover Institution on War, Revolution, and Peace, at Stanford. In 1935, quantum mechanics was still a new discipline, the key to understanding the atom, and Teller was one of perhaps a hundred theoretical physicists in the world who were steeped in the subject. Already he had done pioneering calculations of the structure and vibrations of molecules.

World events in both politics and science, however, were soon to drag Teller out of the classroom. In 1938, Hitler invaded Austria. Later that year, two German chemists, Otto Hahn and Fritz Strassmann, found evidence that the uranium nucleus could be cloven in two, or "fissioned," by the impact of a diminutive neutron, releasing in the process roughly ten million times more energy per gram of material than any chemical reaction. Indeed, it was later realized that a certain rare type, or isotope, of the uranium nucleus, called U-235, is extremely unstable, like a cocked mousetrap.

Shortly after Hahn and Strassmann's discovery, Leo Szilárd, another Hungarian physicist and a close friend of Teller's, suggested that if the fission of one uranium nucleus shook loose several free neutrons in addition to the two large fission halves, then each of these new neutrons could fission another uranium nucleus nearby, and so on, setting in motion an explosive "chain reaction." Szilárd himself began an experiment to investigate the matter. One evening a month later, while Teller was playing a Mozart sonata, he received a long-distance telephone call from Szilárd: "I found the neutrons."

"When I returned to the piano," Teller writes, "I knew that the world might change in a radical manner."

Soon President Roosevelt received the famous letter from Einstein warning him of the possibility of an atomic bomb, and, furthermore, of an atomic bomb in the hands of the Germans. Teller and other prominent physicists were asked to form a presidential advisory committee to explore the feasibility of such a bomb. Some of the formidable problems that had to be solved included (1) separating U-235, the rare isotope of uranium that would make a bomb, from the much more prevalent isotope of uranium found in nature, U-238, and (2) designing a way that elements of the bomb could be brought together quickly, making a "critical mass" of uranium.

The concept of a critical mass in an atomic explosion can be explained as follows: If a solid sphere of uranium atoms is too small, a neutron released in the middle of the sphere will reach the outer surface before it collides with a uranium nucleus and thus harmlessly escape without causing a fission. Without a fission, no chain reaction can get going. The critical mass is the minimum mass of uranium that is needed before a typical neutron emitted near the center of the sphere will likely collide with a uranium nucleus before reaching the outer surface.

In building an atomic bomb, one cannot begin with a sphere of U-235 larger than the critical mass. Since there are always stray neutrons going through space, the bomb would be detonated prematurely in the workshop. Thus, the design problem is to begin with two or more "subcritical" masses, safely separated from each other, and then ram them together to create a critical mass at the desired moment. The critical mass for U-235 at achievable densities is roughly five pounds (the exact number is still classified), corresponding to the explosive energy that would be released by about 25,000 tons of a chemical high explosive like TNT. The atomic bomb that destroyed Hiroshima, an inefficient bomb by later standards, released the energy of about 10,000 tons of TNT.

THREE

IN THE FALL OF 1940, when the concept of the atomic bomb was still being debated, Enrico Fermi made a fateful comment to Teller. During a walk to the physics laboratory at Columbia University, Fermi asked Teller whether the extreme heat from an atomic bomb could not cause hydrogen atoms to fuse together, releasing yet another new source of energy. It was already known that ordinary hydrogen atoms in the sun fuse together slowly to make helium, and this slow and steady reaction, called a thermonuclear reaction, provides the energy of the sun and many other stars.

Fermi's suggestion, in particular, was that the more rare isotope of hydrogen atoms, called deuterium, might fuse together far more rapidly, creating a bomb rather than a mere star. An atomic bomb, near or inside the deuterium, would be the "trigger." A hydrogen bomb (also called a thermonuclear bomb or a fusion bomb) would differ greatly from a uranium bomb (also called an atomic bomb or a fission bomb). Within the tiny atomic nucleus two fundamental forces do battle against each other: the repulsive electrical force between protons and the attractive nuclear force between protons and neutrons. The protons, repelling one another, are like compressed springs waiting to be released, and only the attractive nuclear force prevents them from flying away from one another at great speed. The first force, which gains the upper hand when a large U-235 nucleus is slightly deformed by collision with a neutron, provides the energy of the atomic bomb. The second force, dominating in small nuclei when the protons and neutrons are brought close enough together, powers the hydrogen bomb. A high heat is required to ignite the hydrogen bomb because the deuterium nuclei (each consisting of one proton and one neutron) must be rammed together so closely that the attractive nuclear force

143

between nuclei can overcome the repulsive electrical force and pull the nuclei together.

Teller became obsessed with Fermi's idea. Unlike a sphere of U-235, which is unstable and therefore must be kept below "critical mass," a sphere of deuterium, no matter how large, can be detonated only by extremely high heat and compression. Thus, no critical mass exists for a hydrogen bomb. Unlike the atomic bomb, a hydrogen bomb can be built without any limitation on size—the Soviets once exploded an H-bomb equivalent to 100 million tons of TNT, or 100 megatons. Furthermore, pound for pound of explosive material, a hydrogen bomb releases ten times as much energy as an atomic bomb. Finally, deuterium is readily available in seawater and far easier to separate from ordinary hydrogen than U-235 is from U-238. In short, an H-bomb would be far more powerful and also cheaper than an A-bomb.

FOUR

IN THE SPRING OF 1943, in the scrub brush and desert terrain of New Mexico, the secret Los Alamos Laboratory was created to pursue the atom bomb at full force, with Robert Oppenheimer named the director. Teller was hoping to be designated the head of the theoretical

division, thirty physicists strong, but Oppenheimer appointed Hans Bethe instead.

Although Teller made several important suggestions about the fission bomb, he refused to help with detailed calculations when Bethe asked him to do so. Teller, at his own request, was formally relieved of further responsibilities. What he did, instead, was to pursue his preoccupation with the hydrogen bomb and, according to Bethe, "spent long hours discussing alternative schemes which he had invented for assembling an atomic bomb or to argue about some remote possibilities why our chief design might fail."

After the end of the war, Leo Szilárd and other leading atomic scientists were not eager to continue in weapons work. Bethe remembers: "We all felt that, like the soldiers, we had done our duty. . . . Moreover, it was not obvious in 1946 that there was any need for a large effort on atomic weapons in peacetime." Fermi and Bohr argued that only politics and international negotiation could counter the danger of nuclear weapons. A few years later, Oppenheimer expressed his opposition to the development of Teller's hydrogen bomb in a report of the General Advisory Committee of the United States Atomic Energy Commission: "We base our recommendation on our belief that the extreme dangers to mankind inherent in the proposal wholly outweigh any military advantage that could come from this development. . . . A super bomb might become a weapon of genocide." Teller felt enormously frustrated that he was not receiving scientific and governmental backing for his project, and that his colleagues had "[lost] their appetite for weapons work."

BETWEEN 1946 AND 1951, Teller embarked upon a personal crusade for the Super, at a time when many physicists were not sure that a hydrogen bomb was even theoretically possible. Deeply suspicious of Oppenheimer's "pacifism" and lack of support, Teller sought alliances with a number of powerful political and scientific figures, including Admiral Lewis Strauss, along with Oppenheimer an inaugural member of the AEC, and the physicist Ernest O. Lawrence, winner of the 1939 Nobel Prize for his invention of the cyclotron.

In this and later crusades, Teller was more than overly confident, and he was sometimes duplicitous. Robert Serber, a protégé of Oppenheimer's, remembers an important conference organized by Teller in 1946 to assess the current state of knowledge of the Super: "It became apparent to me," Serber said, "that at every point they were making the most optimistic assumptions and that no solid calculations had really been carried through." Serber went to Teller and suggested that they "tone down some of the more outrageous statements." Teller agreed and the two physicists rewrote the report. As Serber recounts, a couple of months later he came across the (classified) final version of the conference report and discovered that "it was Edward's original report, with all the changes we had agreed on left out."

When the Polish mathematician and physicist Stanislaus Ulam demonstrated in 1950 that Teller's initial design was flawed, Teller was, in Ulam's words, "pale with fury." Remarkably, less than a year later, Ulam and Teller came up with a new concept that worked. Teller has never given credit for this to Ulam. (Ulam's particular and essential contribution, acknowledged by physicists on the scene, was that extreme compression of the deuterium would solve some of the difficulties that had been encountered, especially the loss of needed heat in the form of radiation.) Teller and Ulam intensely disliked each other. In *Memoirs* Teller writes scornfully that "Ulam didn't understand my new design and claimed it would never work."

Teller's lifelong denial of Ulam's contribution, while far from forgivable, may at least be partly explained by Teller's extreme commitment to the fusion bomb over many years—reflecting a personal passion, emotional involvement, and sense of ownership that is common among scientists. These qualities, coupled with Teller's brooding egotism, were probably dominating factors in his quest for the hydrogen bomb. No doubt his fear of the Russians, his sense of duty, and his belief that peace could be achieved only through powerful weapons were sincere and genuine. But his personal ambition seems to have been even stronger. In an unguarded and self-revealing moment in *Memoirs,* Teller says of the mathematician John von Neumann, his close friend and fellow Hungarian: "Like the other scientists from Hungary that he knew well, he had only one ambition, and that was to see his ideas succeed."

DESPITE INCREASING enthusiasm for Teller and Ulam's breakthrough idea for the fusion bomb, Teller himself felt that the only solution to what he considered an unfriendly and timid atmosphere at Los Alamos was to create a second, rival weapons laboratory—which he, of course, would control. In *Memoirs,* Teller denies campaigning for the new laboratory before he left Los Alamos in October 1951: "It would be unseemly to advocate a second laboratory while working at Los Alamos." However, according to an entry from the diary of Gordon Dean, then chairman of the powerful AEC, Teller met in Washington with Dean on April 4, 1951, and pleaded for a new weapons laboratory, to be staffed with 50 senior scientists, 82 junior scientists, and 228 assistants.

One can imagine the commanding presence of Teller in this and many other meetings behind the scenes or on such important committees as the AEC's Reactor Safeguard Committee, Nelson Rockefeller's Commission on Critical Choices for Americans, and Reagan's White House Science Council. In addition to his imposing profile, his heavy-lidded eyes and thick, bushy eyebrows, Teller's voice, in the words of *The New York Times* journalist William Broad, "could easily be called doom-laden, especially when he set out his words one by one, like great blocks of granite." Teller could also be cutting. When Teller was given the presidential award of the Fermi Prize in 1962, President Kennedy asked the physicist about Teller's proposed plan to use nuclear weapons to blast a second Panama Canal. Teller replied: "It will take less time to complete the canal than for you to make up your mind to build it."

With the crucial help of Ernest Lawrence at the University of California at Berkeley and of the AEC, Teller's new weapons laboratory was established at Livermore, California, in 1952. Eventually, the Livermore Laboratory, with thousands of workers, would be as formidable as Los Alamos; it trained many of the nation's leading young weapons scientists and established an innovative Department of Applied Science, in conjunction with the University of California at Davis, for the education of graduate students. Teller, who looks back

with avuncular fondness on the young weapons physicists he trained at Livermore, writes that "of all the
things I have done in my life, I am most proud of my
role in the establishment and work of the Livermore
laboratory."

IF THE LIVERMORE LAB was Teller's proudest achievement, his testimony at the Oppenheimer hearings in
1954 was his most painful mistake, one that he now
says he regrets. The affair began in late 1953 when William Borden, formerly director of a committee of the
AEC, wrote to FBI Director J. Edgar Hoover that
Oppenheimer was either a Communist agent himself or
supported Communist agents and could not be trusted
with military secrets. Hoover sent Borden's letter to
President Eisenhower.

Teller's own view of Oppenheimer, spelled out in FBI
interviews and other records at the time, was that
"Oppie" had misadvised the government on weapons
development, particularly in opposing the H-bomb and
the creation of the Livermore lab, but that in no way
was Oppenheimer subversive or disloyal. Teller faults
Eisenhower for ordering the witch-hunt-style hearings
and writes, "In retrospect, I should have said at the
beginning of my testimony that the hearing was a dirty
business, and that I wouldn't talk to anyone about it."

In fact, Teller testified that he "would like to see the vital interests of this country in hands which I understand better, and therefore trust more."

In *Memoirs*, Teller gives a convoluted and deceptive explanation of the circumstances and motives behind his testimony. Perhaps the most reliable guide is to be found in the notes of a meeting Teller had, six days before his testimony, with Charter Heslep of the United States Information Service. In a report of that meeting to Lewis Strauss, Heslep writes that Teller

> regrets the case is on a security basis because he feels that it is untenable. . . . Since the case is being heard on a security basis, Teller wonders if some way can be found to "deepen the charges" to include a documentation of the "consistently bad advice" that Oppenheimer has given.

This report clearly suggests that Teller had decided that he would use the hearings, unjustified as they were, as an instrument to dethrone a powerful enemy who had opposed him since his first days at Los Alamos and attempted to foil many of his projects.

Beyond Teller's acknowledgment in *Memoirs* that he was "stupid" and wrong to testify in a hearing that should never have taken place, beyond his anguish over the resulting loss of colleagues and friends, one senses that he was in a state of inner conflict about the right thing to do. To be sure, Teller often used brutal tactics to advance his ambitions; but he was also acutely aware of standards of fair play and of the repressive methods of totalitarian governments. Other powerful figures in Washington, such as Strauss, wanted to silence Oppenheimer; they put pressure on Teller to give damaging testimony so that Oppenheimer would be stripped of his security clearance. In the Oppenheimer affair, it can be argued that Teller was, at least to some degree, a victim as well as a victimizer.

ONE IS STRUCK by how important Teller's friends were to him. Some of them, including Leo Szilárd, John von Neumann, Enrico Fermi, Ernest Lawrence, John Wheeler, and Maria Mayer—all leading physicists or mathematicians—remained close to him during and after the Oppenheimer affair. *Memoirs* quotes from a dozen candid and personal letters from Teller to Mayer. Of Fermi, Teller says, he "didn't care whether I was right or wrong—he simply wanted to heal the schism."

Shortly after the Oppenheimer hearings, at the young age of fifty-three, Fermi was diagnosed with terminal cancer. Teller flew at once to his bedside, in Chicago. According to Teller's account, instead of talking about himself, Fermi asked about Teller, knowing that he was under harsh attack. Teller writes that the meeting, his last visit with Fermi, "was characteristic of Fermi's generous nature. He had almost all the good traits that a friend could have . . . open-hearted, good-humored, and alert to others' needs."

I recently visited one of Teller's old friends, George Hatsopoulos, a highly accomplished, seventy-five-year-old scientist and businessman who in 1956 founded the Thermo Electron Company. Hatsopoulos told me how fond he was of Teller, who had, he said, a great sense of humor, not always common among scientists. "With me," Hatsopoulos said, "he was always an uncomplicated person, a genuine friend. As soon as a senator or someone like that would walk into the room, Edward's eyes would light up. I could see the change. He wanted to impress the person. But he was honest with me. . . . Edward is brilliant, but naive."

Teller emerges from such accounts and from his own book as capable of being both a good friend and a feared enemy, uncomplicated and genuine at one moment, a salesman driven to impress at the next—combative and vulnerable, slyly political and naive, honest and dishonest. And, now and then, aware of these contradictory qualities.

FIVE

IN THE EARLY 1980S, well after he became a professor emeritus at the University of California, well after he had retired from the Livermore Laboratory, Teller became involved in the last major controversy of his life: the Strategic Defense Initiative (SDI), popularly called Star Wars. For decades, Teller had been keenly interested in defense against nuclear bombs, particularly bombs mounted on incoming missiles. The idea of the "X-ray laser"—by which a nuclear bomb set off in space could power an intense X-ray laser beam that could quickly focus on and destroy missiles shortly after launch—had excited some scientists at Livermore. Teller strongly supported the concept and proposed it to his friend President Reagan. When Reagan asked the enthusiastic Teller if such an antimissile system could work, Teller replied, "We have good evidence that it would." Reagan then made his famous SDI speech of March 23, 1983, with the implied commitment of billions of dollars.

In fact, there was little scientific evidence that the X-ray laser would work, and that evidence was based on crude and inconclusive experiments. It is still unclear whether such a system is scientifically possible, and, even if it is, many analysts believe that the X-ray laser could never be part of a workable defense system. After interviewing a number of weapons experts at Livermore, William Broad wrote in his book *Teller's War* that the proposed program was "carved out of thin air over the objections of key Livermore officials." One of those Livermore scientists, Hugh DeWitt, told me that he and other scientists considered the X-ray laser to be "utter nonsense."

THE SDI EPISODE is part of a continuing pattern in Teller's career of misrepresenting scientific facts in order to get what he wanted. Unfortunately, Teller's methods have been adopted by many younger scientists and officials in the national weapons industry, which is dependent on Congress for funding. The latest example of such deception is the proposal, by scientists and weapons analysts at Los Alamos and Livermore, of so-called mini-nukes, very-low-yield nuclear weapons that can allegedly burrow into hardened bunkers and explode deep enough underground so that no significant radioactivity is released on the surface. Such hypothetically "clean" nuclear weapons could be used in a variety of tactical ways in civilian areas. However, an analysis by the Federation of Atomic Scientists suggests that these claims are impossible. Unable to burrow deep enough without destroying their own hardware, such weapons would throw up a massive crater of radioactive dirt and other radiation, lethally contaminating the surrounding environment.

The decisive responsibility for evaluating these claims lies with Congress. But too often in budget hearings legislators submissively defer to weapons scientists. In some ways, scientists at the national weapons laboratories, necessary as they are for national defense, should be viewed not as scientists but as salespeople, trying to sell their products. Accordingly, their claims should be carefully analyzed by independent scientists and organizations.

In 1962, Teller wrote, "In a dangerous world we cannot have peace unless we are strong." Such a military credo is certainly defensible; but in the hands of Teller and others, it has led to exaggerated claims about the abilities of particular weapons; furthermore, it has been used to argue not just for "sufficient" military strength but also for maximum and superior strength, without any limitation of arms. As many analysts have pointed out, we could do much better in achieving security at manageable cost with a smaller force, sufficient to counter conceivable dangers, regulated under arms control agreements. By and large, the few arms agreements we have had with the former Soviet Union have been successful.

When one reads through Teller's many statements and considers his consistent and inflexible opposition to arms control agreements despite drastic geopolitical changes, one suspects that his real interest lies not in sufficient military strength but in world supremacy. Such global dominance might protect the United States militarily, but it would exclude Americans from citizenship in the world. It would exclude real cooperation with and understanding of other nations and cultures. Paradoxically, it may actually increase the likelihood of the kind of attack that happened on September 11, 2001.

AT A RECEPTION for Soviet premier Mikhail Gorbachev at the White House on December 8, 1987, President Reagan introduced Teller to Gorbachev, saying, "This is Doctor Teller." When Teller reached out his hand, Gorbachev stood frozen and silent. Reagan then added: "This is the famous Doctor Teller." Still without shaking hands, Gorbachev said: "There are many Tellers." Indeed, there are.

(2002)

DARK MATTER

IN THE FALL OF 1950, Vera Rubin, then a graduate student in astronomy at Cornell, decided that she wanted to present her master's thesis work to the December meeting of the American Astronomical Society. Rubin had analyzed the motions of about 110 galaxies—all the galaxies whose velocities were reasonably known. The galaxies, each an island of billions of stars surrounded by the black sea of space, were located as far as ten million light-years from the Milky Way, our own galaxy.

That the universe is expanding, with all the galaxies flying away from one another, had already been established. Rubin found that the galaxies also appeared to be moving in circles, like horses on a merry-go-round. From her cramped student's office in Ithaca, New York, she submitted the title of her proposed presentation: "The Rotation of the Universe." At the time, Rubin was twenty-two years old. She was also pregnant with her first child.

The chairman of the astronomy department at Cornell told the swelling young woman that, as she was going to have a child any moment and wouldn't be able to go to the meeting, he would give her talk for her. Furthermore, as she was not a member of the American Astronomical Society, his name would appear on the paper. At which point, Mrs. Rubin spoke up, saying, "Oh, I can go."

Rubin's son was born at the end of November. She and her husband didn't have a car, so her parents drove them and the newborn to the meeting in Haverford, Pennsylvania, an anguished trip through the snow while Rubin was nursing her child in the backseat. In Haverford, she walked into the meeting, gave a ten-minute talk to the national assembly of scientists, and left.

The astronomers were appalled by Rubin and her paper on "The Rotation of the Universe." And, indeed, better data in later years negated the bold proposal of a cosmic merry-go-round. However, Rubin was proven correct in her claim that galaxies have large sideways motions that depart from the overall expansion of the universe, a finding that continues to challenge the simple picture of the Big Bang. In such work, she was a world pioneer.

If Rubin had known in advance all the objections to her work—that such contrary motions couldn't exist, that there were not enough galaxies to draw her conclusions, that the distances and brightnesses were not known well enough—she might never have undertaken the project. But she was far removed from the traditional powerhouses of science at Princeton and Harvard and Caltech. She knew only one or two professional astronomers, and she was twenty-two years old. "The only motivation that I can point to is just plain old curiosity," she told me in an interview decades later.

> Curiosity really has motivated an enormous amount of my work. . . . I really tend to work pretty much alone. I personally don't often interact with theorists. . . . I remember, in the early [years] of the large-scale motion problems, some of the people I admire most telling me that you can't have large-scale motions because any irregularity since the early universe would be damped out. I mean they gave me all these reasons, which impressed me—really terrified me—as to why you couldn't have large-scale motions. But if you ultimately get to the point where . . . that is what the observations show, then the theorists just have to tell themselves that they have missed something.

Rubin, a self-deprecating woman now with cropped white hair and a grandmotherly smile, has always had a penchant for working alone, for avoiding the pack, for finding interesting problems that no one else is pursuing.

She has never enjoyed controversy. The negative reactions to her master's thesis work sufficiently frightened her that she abandoned those studies for years. Instead, she sniffed out other astronomical problems where she could work with minimum disturbance.

Her sense of smell has been excellent. In the late 1960s, she began measuring the motion of gas in individual galaxies, a seemingly mundane topic. From those studies, she eventually made her greatest discovery: that more than eighty percent of the mass of spiral galaxies is invisible—the so-called dark matter. Dark matter is some kind of mass that doesn't emit any light. We still don't know what it is, but dark matter accounts for the majority of the material stuff of the universe. Dark matter dominates the gravity of a galaxy. Dark matter is the invisible elephant in the room.

Vera Rubin is a woman in love with astronomy. When she talks about her work, her voice sounds as if she were telling a good story about one of her children. "For me, doing astronomy is incredibly great fun," she says. "It's just an incredible joy to get up every morning and come to work." Over time, those joyful mornings have brought international acclaim. In 1988, Rubin was awarded an honorary doctorate from Harvard, to be followed by other honorary degrees. In 1993, she received the National Medal of Science, America's highest scientific award. In 1996, she won the Gold Medal of the British Royal Astronomical Society, the first woman to receive that award since Caroline Herschel, in 1828. In 2003, she was awarded the Gruber International Cosmology Prize, in 2004 the National Academy of Sciences Watson Prize.

Just before the twenty-two-year-old Rubin fled from the conference hall of the 1950 meeting of the American Astronomical Society, the editor of the *Astronomical Journal* caught up with her and explained that he could never publish a paper titled "The Rotation of the Universe." So the title was changed to the somewhat more modest "Differential Rotation of the Inner Metagalaxy." For years afterward, Rubin received requests for her thesis.

VERA RUBIN was born in July 1928, the second daughter of Philip and Rose Cooper. Philip was an electrical engineer. "He had a very analytical way of looking at things," Rubin recalls. "I enjoyed that very much." As a child, growing up in the Washington, D.C., area, Rubin had a bed placed beneath north-facing windows. She became obsessed with watching the stars move through the night. "By about age twelve," she says, "I would prefer to stay up and watch the stars than go to sleep. There was just nothing as interesting in my life as watching the stars every night." During the occasional meteor showers, young Rubin would memorize where each object went so that in the morning she could make maps. Soon, Rubin decided on her future profession. "I knew I wanted to be an astronomer," she recalls. "I didn't know a single astronomer, but I just knew that was what I wanted to do."

In high school, Rubin set her sights on Vassar. She had read about Maria Mitchell, discoverer of a comet in 1847 and the first highly recognized female scientist in America. When Vassar was opened in 1865, Mitchell became its first professor of astronomy. Almost a century later, there were still not many colleges where a woman could study astronomy. And even Vassar had only a tiny department.

The day that Rubin learned she'd gotten a scholarship to Vassar, she met her high-school physics teacher in the hallway and excitedly told him the news. "He was a real macho guy," remembers Rubin. "I really don't think he knew how to relate to a young girl in his class, and it became a terrible battle of wills." As she vividly remembers, he looked at her, with her acceptance letter in hand, and said that as long as she stayed away from science she would do satisfactorily. Says Rubin, "It takes an enormous self-esteem to listen to things like that and not be demolished."

In the last year of her studies at Vassar, Rubin sent a postcard off to Princeton, asking them for a catalogue of their graduate school. The dean of the graduate school replied that since they didn't accept women, he wouldn't be sending her the catalogue. So Rubin went to Cornell for her master's and then on to Georgetown for her Ph.D., where she wrote her thesis under the distinguished physicist and cosmologist George Gamow.

Rubin was not demolished. Besides attending to her own deepening career in science, she began creating new scientists. Her second child was born in 1952, her third in 1956, her fourth in 1960. In her curriculum vitae, Rubin proudly lists all of her children, each with his or her own Ph.D.: David's and Allan's in geology, Judith's in cosmic ray physics, Karl's in mathematics. Rubin's husband, Robert, whom she credits with providing strong support through her challenging career, is a mathematician and biologist.

Against all odds, Rubin has managed to be a mother and wife as well as an outstanding scientist. Few of the great women scientists of the twentieth century ever married. One thinks of the astronomer Henrietta Leavitt, the nuclear physicist Lise Meitner, the X-ray crystallographer Rosalind Franklin, the geneticist Barbara McClintock. Fewer still have raised families.

IN ONE WAY or another, much of Rubin's work has dealt with the measurements of cosmic mass. In astronomy, mass is important. Mass produces gravity. And gravity in turn determines the structure of galaxies, the manner in which galaxies group and move about in space, and the very nature of space itself.

A difficult aspect of astronomy, as opposed to all other sciences, is that the objects of study are beyond human reach and control. A galaxy cannot be tilted for a better view, a cosmic magnetic field cannot be increased by turning a knob in the lab. By necessity, astronomy is a science of inferences. In the case of measuring cosmic mass, for example, the stars and galaxies cannot be pulled down to earth and put on a scale. Their mass must be estimated indirectly.

There are two principal methods. The first method measures the light produced by the mass and then makes assumptions about how much mass is needed to produce that much light. For example, we know well the mass and luminosity of our own sun. Thus, if we assume that a galaxy is made up of stars like our sun, or stars whose properties relative to our sun are well known, then measuring the total light tells us the total mass.

The second method infers the amount of mass in a region by its gravitational effects. For example, there is a well-defined relation between the mass of a galaxy, the distance of a star from the center of that galaxy, and the velocity of that star. If an astronomer measures the distance and velocity of an orbiting star or a globule of gas, he or she can deduce the mass pulling on that object. (A third method, called "gravitational lensing" and actually a version of the second, measures mass by the amount that it deflects passing light rays.)

The first method determines mass we can see but, by definition, not mass we cannot see. The second determines both mass that is seen and mass that is unseen but obviously provides few clues about what we cannot see. When the two methods agree, then all of the mass is in light-producing stuff, like ordinary stars. When the second method gives a larger answer than the first, then some of the mass must be invisible.

IN 1933, the eccentric but brilliant Swiss American astronomer Fritz Zwicky used the two methods to estimate the amount of mass in a "cluster" of galaxies, a group of galaxies that seem to be orbiting one another under their mutual gravitational attraction. A single galaxy is typically about a hundred thousand light-years in diameter, where a light-year is the distance that light travels in one year. (The sun is about eight light-minutes away; the nearest star is a few light-years away.) A cluster of galaxies might have anywhere from ten to a thousand galaxies and extend over a distance of a few million to twenty million light-years. (Zwicky's first cluster had only seven galaxies.) Zwicky concluded that if a cluster of galaxies was to remain gravitationally bound together, without flying apart, then there had to be a great deal of invisible mass—the first evidence for dark matter. However, Zwicky's measurements were extremely approximate. Furthermore, he dealt with groups of galaxies rather than individual galaxies, where the behavior is much better understood.

Vera Rubin began measuring the masses of individual galaxies in 1970. What she actually measured were the colors of glowing atoms of hydrogen and helium orbiting the centers of galaxies at various distances from the center. Because motion causes the color of a moving source of light to shift in a regular way, just as the pitch of a moving source of sound shifts, one can infer the velocity of the gas from its shifted color. The velocity and distance then give the amount of mass pulling on the particle of gas, following the second technique discussed above.

To measure the colors of distant and thus faint material, Rubin collaborated with a master instrument maker, Kent Ford, who had built the best "spectrograph" in the business. At the time, both Rubin and Ford worked at the Carnegie Institution of Washington, a private institution founded by Andrew Carnegie in 1902 to pursue basic knowledge in the sciences. Rubin had joined the institution in 1965. Rubin and her colleagues attached Ford's spectrograph to telescopes at the

Kitt Peak National Observatory, in Arizona, at the Lowell Observatory, in Flagstaff, Arizona, and at the Cerro Tololo Inter-American Observatory, in Chile. The dim light from the galactic source passed through the spectrograph, where it was split into its various colors as if going through a prism, multiplied by a factor of ten by a "Carnegie" image tube, and finally recorded by photographic emulsion. (Today, everything is digital, without film.)

Rubin says that she chose what would become her most famous line of research because Ford's instrument was newly available, because she had always been interested in "galaxy dynamics," and because it was a "valuable [project] to do, but one that was not so in the forefront of astronomy that everyone was doing it." She was also, she says, interested in "what happens at the edges of galaxies."

It had long been known that the luminosity of a typical galaxy is very intense at the center of the galaxy and decreases rapidly as one moves away from the center. Since a galaxy's luminosity is produced mostly by stars, the obvious inference is that the stars of a galaxy are concentrated and dense at the center and diffuse farther out, like the population of a big city. If most of the mass of a galaxy is in the form of stars, then the mass should be concentrated at the center. In which case one would expect from the law of gravity that the velocity of an orbiting particle of gas would decrease with increasing distance from the center, as the gravitational attraction to the center gets weaker and weaker.

What Rubin and her collaborators found, instead, was that the velocities of orbiting particles did not decrease at all. Rather, the velocities continue at a constant value of a couple hundred kilometers per second (about a half million miles per hour) outward from the center of the galaxy, and continue outward without decrease far beyond the region lit up by stars. From this result, one infers that the mass of the galaxy also increases outward far beyond the visible light. In fact, there seems to be no outer edge to the mass of galaxies. Evidently, some giant halo of invisible material surrounds the center of each galaxy.

Despite Zwicky's work in 1933 and some theoretical proposals in the early 1970s that massive haloes were needed to stabilize the rotations of spiral galaxies, Rubin's finding of dark matter was an unwanted surprise. As Rubin recalls, "I think many people initially wished that you didn't need dark matter. It was not a concept that people embraced enthusiastically." As in her earlier work on the bulk motions of whole galaxies, dark matter would require a revision in cosmological thinking. Rubin, a woman who dislikes controversies, always seems to be causing trouble. "If there were no problems," she says in her quietly subversive way, "it wouldn't be much fun."

Today, astronomers believe that only about five percent of the mass of the universe exists in the form of ordinary, light-producing material. Another twenty-five percent is believed to be "dark matter," as yet unidentified. And about seventy percent exists in an even stranger, nonmaterial form called "dark energy," which actually exerts a repulsive gravitational force.

A PHOTOGRAPH of Vera Rubin from 1965 shows her standing beneath the big telescope at the Lowell Observatory, in Flagstaff, Arizona. To her left is a giant wheel for counterbalancing the telescope and rotating its gaze through the heavens. Rubin, a tiny figure beneath the behemoth astronomical instruments, wears a skirt and a sleeveless blouse and looks as if she is about to go to lunch rather than investigate the secrets of the universe.

In 1965, Rubin "integrated" the Palomar Observatory, in San Diego, then possessing the largest telescope in the world, by being the first woman to work there. She has become a role model for women in science. Such examples are badly needed. If one counts the Nobel Prizes in science awarded from 1963 to 2003—taking 1963 as the beginning of the modern women's movement with Betty Friedan's *The Feminine Mystique*—one finds that 261 men have received the prize and only 7 women. In the physical sciences, female graduate students increased from twenty-three percent in 1990 to twenty-nine percent in 1999.

Rubin, now seventy-five years old, still works at her office at the Department of Terrestrial Magnetism at the Carnegie Institution, studying the spectra of galaxies. She still makes discoveries. In the early 1990s, she found that different groups of stars in the obscure galaxy NGC 4550 orbit in opposite directions, a still baffling phenomenon. About once a month, a different high-school girl and her father visit Rubin to talk about careers for women in science. Sixth-grade girls send her videos of their classroom projects impersonating a famous scientist, Vera Rubin. In the videos, the girls wear long dresses that drag on the floor, hats, pocketbooks, and Rubin's trademark large-rimmed glasses.

But impressions about what is possible in life and what is not start at an even younger age. Rubin tells the story of when her then three-year-old granddaughter discovered that her toy rabbit was sick. A visiting uncle said to the little girl, "Well, you be the doctor and I'll be the nurse, and we'll fix it." To which the granddaughter objected, saying, "Boys can't be girls." By the age of three, the child already believed that only men were doctors and only women were nurses.

"So you may talk about role models and your thinking about colleges," says Rubin, "but this happens at the age of three. I think it's a terrible problem. It sets in very young. Somehow or other, you have to raise little girls who have enough confidence in themselves to be different."

(2004)

A SCIENTIST DYING YOUNG

THE LIMBER YEARS for scientists, as for athletes, generally come at a young age. Isaac Newton was in his early twenties when he discovered the law of gravity, Albert Einstein was twenty-six when he formulated special relativity, and James Clerk Maxwell had polished off electromagnetic theory and retired to the country by thirty-five. When recently I hit thirty-five myself, I went through the unpleasant but irresistible exercise of summing up my career in physics. By this age, or another few years, the most creative achievements are finished and visible. You've either got the stuff and used it or you haven't.

In my own case, as with the majority of my colleagues, I concluded that my work was respectable but not brilliant. Very well. Unfortunately, I now have to decide what to do with the rest of my life. My thirty-five-year-old friends who are attorneys and physicians and businessmen are still climbing toward their peaks, perhaps fifteen years up the road, and are blissfully uncertain of how high they'll reach. It is an awful thing, at such an age, to fully grasp one's limitations.

Why do scientists peak sooner than most other professionals? No one knows for sure. I suspect it has something to do with the single focus and detachment of the subject. A handiness for visualizing in six dimensions or for abstracting the motion of a pendulum favors a nimble mind but apparently has little to do with anything else. In contrast, the arts and humanities require experience with life, experience that accumulates and deepens with age. In science, you're ultimately trying to connect with the clean logic of mathematics and the physical world; in the humanities, with people. Even within science itself, a telling trend is evident. Progressing from the more pure and self-contained of sciences to the less tidy, the seminal contributions spring forth later and later in life. The average age of election to England's Royal Society is lowest in mathematics. In physics, the average age at which Nobel Prize winners do their prize-winning work is thirty-six; in chemistry it is thirty-nine, and so on.

Another factor is the enormous pressure to take on administrative and advisory tasks, which descends on you in your midthirties and leaves time for little else. Such pressures also occur in other professions, of course, but it seems to me they arrive sooner in a discipline where talent flowers in relative youth. Although the politics of science demands its own brand of talent, the ultimate source of approval—and invitation to supervise—is your personal contribution to the subject itself. As in so many other professions, the administrative and political plums conferred in recognition of past achievements can crush future ones. These plums may be politely refused, but perhaps the temptation to accept beckons more strongly when you're not constantly pushing off into new research.

Some of my colleagues brood as I do over this passage, many are oblivious to it, and many sail happily ahead into administration and teaching, without looking back. Service on national advisory panels, for example, benefits the professional community and nation at large, allowing senior scientists to share with society their technical knowledge. Writing textbooks can be satisfying and provides the soil that allows new ideas to take root. Most people also try to keep their hands in research, in some form or another. A favorite way is to gradually surround oneself with a large group of disciples, nourishing the imaginative youngsters with wisdom and perhaps enjoying the authority. Scientists with charisma and leadership contribute a great deal in this manner. Another, more subtle tactic is to hold on to the reins, single-handedly, but find thinner and thinner horses to ride. (This can easily be done by narrowing one's field in order to remain "the world's expert.") Or simply plow ahead with research as in earlier years, aware or not that the light has dimmed. The one percent of scientists who have truly illuminated their subject can continue in this manner, to good effect, well beyond their prime.

For me, none of these activities offers an agreeable way out. I hold no illusions about my own achievements in science, but I've had my moments, and I know what it feels like to unravel a mystery no one has understood before, sitting alone at my desk with only pencil and paper and wondering how it happened. That magic cannot be replaced. When I directed an astrophysics conference one summer and realized that most of the exciting research was being reported by ambitious young people in their midtwenties, waving their calculations and ideas in the air and scarcely slowing down to acknowledge their predecessors, I would have instantly traded my position for theirs. It is the creative element of my profession, not the exposition or administration, that sets me on fire. In this regard, I side with the great mathematician G. H. Hardy, who wrote (at age sixty-three) that "the function of a mathematician is to do something, to prove new theorems, to add to mathematics, and not to talk about what he or other mathematicians have done."

IN CHILDHOOD, I used to lie in bed at night and fantasize about different things I might do with my life, whether I would be this or that, and what was so delicious was the limitless potential, the years shimmering ahead in unpredictability. It is the loss of that I grieve. In a way, I have gotten an unwanted glimpse of my mortality. The private discoveries of new territory are not as frequent now. Knowing this, I might make myself useful in other ways. But another thirty-five years of supervising students, serving on committees, reviewing others' work, is somehow too social.

Inevitably, we must all reach our personal limits in whatever professions we choose. In science, this confrontation happens at an unreasonably young age, with a lot of life remaining. Some of my older colleagues, having passed through this soul-searching period themselves, tell me I'll get over it in time. I wonder how. None of my fragile childhood dreams, my parents' ambitious encouragement, my education at all the best schools, prepared me for this early seniority, this stiffening at age thirty-five.

(1984)

PORTRAIT OF THE WRITER AS A YOUNG SCIENTIST

WHEN I TURNED thirty-five, I wrote an essay for *The New York Times Magazine* about my distressing awareness that I would soon be an old man in my field. That field was theoretical physics, where people do their best work at a famously young age. Now, sixteen years later, having long since given up physics for a profession in which I am still young, I find myself looking back on my life as a scientist and what I so miss.

I miss the purity. Theoretical physicists, and many other kinds of scientists, work in a world of the mind. It is a mathematical world without bodies, without people, without the vagaries of human emotion. A physicist can imagine a weight hung from a spring bouncing up and down and fix this mental image with an equation. If friction with air becomes an unwanted nuisance, just imagine the weight in a vacuum.

Much of science, in fact, is built on these pure pictures of the mind. And the equations themselves are beautiful. The equations have a precision and elegance, a magnificent serenity, an indisputable rightness. I remember so often finding a sweet comfort in my equations after arguing with my wife about this or that domestic concern or fretting over some difficult decision in my life or feeling confused by a person I'd met. I miss that purity, that calm.

Of course, other occupations also deal with ideas. But the ideas are often complicated with the ambiguity of human nature. The exquisite contradictions and uncertainties of the human heart do indeed make life interesting; they are why God held the apple in front of Eve and then forbade her to eat it, they inspire artists and art, they are why the poet Rainer Maria Rilke wrote that we should try to love the questions themselves.

All that is necessary and good. But I miss the answers. I miss the rooms I could enter, the language that sounded clear as a struck bell.

I miss the exhilaration of seeing brilliant people at work, watching their minds leap right in front of me, not the brooding intelligence of writers, but an immediate mental agility, pole vaults and somersaults and triple axels on the ice. Richard Feynman once walked into my tiny office at Caltech and, in twenty minutes at the blackboard, outlined the basic equations for the quantum evaporation of spinning black holes, an ingenious idea that had just occurred to him on the spot. When I was stymied by a tough astrophysics problem at Cornell, the great theoretician Edwin Salpeter, while lying on the floor of his living room with back pain, instantly drew an analogy between the slow drift of stars orbiting a disruptive mass and the random motion of a drunk stumbling around on a street with an open sewer hole.

Others I watched from the front row: the British astrophysicist and astronomer royal Martin Rees, the Nobel Prize–winning particle physicist Steven Weinberg. In the presence of these minds I felt humbled as well as excited. I miss the humility; it made me crouch down and observe. I listened more than I talked. I took in.

Most of all, I miss the intensity. I miss being grabbed by a science problem so that I could think of nothing else, consumed by it during the day and then through the night, hunched over the kitchen table with my pencil and pad of white paper while the dark world slept, tireless, electrified, working on until daylight and beyond.

Every creative field has its moment of inspiration, the struggle to that moment and then the surge of insight. In most occupations, the aftermath is a slow working-out of the idea. As a writer, even when I am writing well, I cannot write more than six hours at a time. After that I am exhausted, and my vision has become clouded by the inherent subtleties and uncertainties of the work. Then I must wait for the words to shift and settle on the page and my own strength to return.

But as a scientist, I could be gripped for days at a

time. I could go for days without stopping. Because I wanted to know the answer. I wanted to know the telltale behavior of matter spiraling into a black hole, or the maximum temperature of a gas of electrons and positrons, or what was left after a cluster of stars had slowly lost mass and drawn in on itself and collapsed.

When in the throes of a new problem, I was driven night and day, compelled because I knew there was a definite answer, I knew that the equations inexorably led to an answer, an answer that had never been known before, an answer waiting for me. That certainty and power and the intensity of effort it causes I dearly miss. It cannot be found in most other professions.

Sometimes, I wonder if what I really miss is my youth. Purity, exhilaration, intensity—these are aspects of the young. In a way, it is not possible at age fifty for me to look back on myself in my twenties and early thirties and understand anything more than the delicious feeling of immortality, the clarity of youth, the feeling that everything was possible.

I do miss my youth. And now, as a writer still finding my stride, while I can reasonably expect another couple of decades to thrive, I know that this second profession too will come to an end, that I will inevitably dwindle down to the physical as well as mental and artistic end, the final end. Of course, I want to be young again.

If given a chance to start over, I would do just what I did, to be not only a young man in the shimmering of youth but a scientist. I would want again to be driven day and night by my research. I would want the beauty and power of the equations. I would want to hear that call of certain truth, that clear note of a struck bell.

(2000)

PRISONER OF THE WIRED WORLD

NOT LONG AGO, while sitting at my desk at home, I suddenly had the horrifying realization that I no longer waste time. It was one of those rare moments when the mind is able to slip out of itself, to gaze down on its convoluted gray mass from above, and to see what it is actually doing. And what I discovered in that flicker of heightened awareness was this: from the instant I open my eyes in the morning until I turn out the lights at night, I am at work on some project. For any available quantity of time during the day, I find a project, indeed I feel compelled to find a project. If I have hours, I can work at my laptop on an article or book. If I have a few minutes, I can answer a letter. With only seconds, I can check telephone messages. Unconsciously, without thinking about it, I have subdivided my waking day into smaller and smaller units of "efficient" time use, until there is no fat left on the bone, no breathing spaces remaining. I rarely goof off. I rarely follow a path that I think might lead to a dead end. I rarely imagine and dream beyond the four walls of a prescribed project. I hardly ever give my mind permission to take a recess, to go outdoors, and play. What have I become? A robot? A cog in a wheel? A unit of efficiency myself?

I can remember a time when I did not live in this way. I can remember those days of my childhood when I would walk home from school by myself and take long detours through the woods. With the silence broken only by the sound of my own footsteps, I would sit on the banks of Cornfield Pond and waste hours watching tadpoles in the shallows or the sway of water grasses in the wind. My mind meandered. I thought about what I wanted for dinner that night, whether God was a man or a woman, whether tadpoles knew they were destined to become frogs, what it would feel like to be dead, what I wanted to be when I became a man, the fresh bruise on my knee. When the light began fading, I wandered home.

I ask myself: What happened to those careless, wasteful hours at the pond? Has the world changed, or just me? Of course, part of the answer, perhaps a large part, is simply that I grew up. Besides the unreasonable nostalgia that most of us have for our youth, adulthood undeniably brings responsibilities and career pressures and a certain consciousness of the weight of life. It is extremely difficult to disentangle the interior, personal experience of aging, strapped with these new burdens, from any change in the exterior world. Yet, I sense that

some enormous transformation has indeed occurred in
the world from the 1950s and '60s of my youth to the
twenty-first century of today. A transformation so vast
that it has altered all that we say and do and think, yet
often in ways so subtle and pervasive that we are hardly
aware of them. Among other things, the world is faster,
less patient, louder, more wired, more public.

Some anecdotal examples: A friend who has been
practicing law for thirty years wrote to me that her
"mental capacity to receive, synthesize, and appropri-
ately complete a legal document has been outpaced by
technology." She says that with the advent of the fax
machine and electronic mail, her clients "want immediate turnaround, even on complex matters," and the
practice of law has been "forever changed from a reasoning profession to a marathon." Another friend who
works at a major software company described to me the
job interview process. An applicant is interviewed independently by several different people on the selection
committee. Afterward, there are no face-to-face meetings of the committee to discuss the applicant. Instead,
each interviewer, within twenty minutes of completing
the interview, must write up his or her impressions and
send them by e-mail to the other members of the committee. If the transmission of judgment isn't completed
within this time frame, that interviewer is out of the
loop. Other business presses on. Or consider time away

from the office. A family that vacations in the same area of Maine where I spend the summer arrives at their rented cabin with sunglasses, beach towels, and canoe paddles. My friends also unpack cell phones and laptops and modems, so that they can stay connected to their workplace throughout the holiday.

Although I cannot document any general conclusions, I believe that these anecdotes represent common experiences. Haven't we all seen people talking on cell phones while dining or riding the train, deadlines and lead times growing shorter and shorter, video screens imposed in the most unexpected of places? All around me, everywhere I go, I feel a sense of urgency, a vague fear of not keeping up with the world, a vague fear of not being plugged in. I feel like the character K in Kafka's *The Trial,* who lived in a world of ubiquitous suspicion and powerful but invisible authorities. Yet there is no real authority here, only a pervasive mentality. I struggle to understand what has happened to the world and to me, why it has happened, and what exactly has been lost.

The dramatic development of technology, especially high-speed communication technologies, has certainly played a major role in shaping the world of today, both for good and for ill. Technology, however, is only a tool. Human hands work the tool. Behind the technology, I believe that our entire way of thinking has changed, our way of being in the world, our social and psychological ethos. The various qualities of this new world are far too complex and broad to be easily categorized, but I will attempt to gather them under the simplistic heading of the "Wired World." Certainly, few people could deny that the new technologies of the Wired World have improved life in many ways. Some of the less agreeable symptoms and features of the Wired World seem to be:

1. An obsession with speed and an accompanying impatience for all that does not move faster and faster. Among the many examples of our accelerating society in James Gleick's recent book *Faster* is the speed of printers. In my childhood, there were people known as typists, who measured their speed in words per minute, perhaps fifty words per minute for a good typist. Authors accepted that they would need to wait weeks to have a manuscript typed. In the 1970s, with the advent of computer printers, speed became gauged in characters per second. A daisy wheel could spit out forty to eighty characters per second, or a single-spaced page every minute, and an author could print an entire manuscript in one day. Ten years later, that same author quickly became discontented with a mere page per minute when the new generation of ink-jet and laser printers could create five pages per minute. When we become accustomed to speed, it is natural to be impatient with slowness.

2. A sense of overload with information and other stimulation. Our computers are not only faster but they store more and more data. The Internet offers an almost infinite amount of information, at easy access. In the face of this avalanche of facts, far more than can be excavated or digested, it becomes easier to confuse information with knowledge. Television screens now are subdivided to show not only the regular program but also, simultaneously, weather information, the latest values of the Dow Jones and Nasdaq indices, and news headlines. Many people have become accustomed to performing several tasks at the same time, such as conducting business on cell phones while driving or walking or eating.

3. A mounting obsession with consumption and material wealth. According to figures from the U.S. Department of Commerce, adjusted for inflation, in 1960, the middle of my childhood, the consumption per person in the United States was $10,700 (in year 2000 dollars). In 2000, it was $24,400, more than double. Researchers have documented that spending and consuming in the United States are higher than anywhere else in the world.

4. Accommodation to the virtual world. The artificial world of the television screen, the computer monitor, and the cell phone has become so familiar that we often substitute it for real experience. Many new technologies

encourage us to hold at a distance the world of immediate, face-to-face contact. Electronic mail, although very useful in some respects, is fundamentally impersonal and anonymous. The sociologist Sherry Turkle, in her book *Life on the Screen: Identity in the Age of the Internet,* discusses how people in "multiuser domains" (MUDs) have created entire artificial communities in cyberspace, escaping for hours at a time their small rooms and meager closets, the relationships or loneliness of their real lives. This increasingly large part of the population refers to real life as "RL," in contrast with "VR," standing for virtual reality.

5. Loss of silence. We have grown accustomed to a constant background of machine noise wherever we are: cars, radios, televisions, fax machines, telephones, and cell phones—buzzes, hums, beeps, clatters, and whines.

6. Loss of privacy. With many of the new communication technologies, we are, in effect, plugged in and connected to the outer world twenty-four hours a day. Individuals are always accessible, always able to access the world around them. Each of us is part of a vast public network of information exchange, communication, and business. This mentality of public connectedness is invisible but always present, like the air.

Aside from the particular technologies, these fundamental qualities of the Wired World have not appeared suddenly or even only during the period since my childhood at Cornfield Pond. They are part of a trend of ever-increasing speed and public access over the last couple of centuries and longer. In recent decades, however, this trend has accelerated to a disturbing degree. If we have indeed lost in some measure the quality of slowness, have lost a digestible rate of information, immediate experience with the real world, silence, and privacy, what exactly have we lost? More narrowly, what have I personally lost when I no longer permit myself to "waste" time? When I never let my mind spin freely, without friction from projects or deadlines, when I never let my mind think about what it wants to think about, when I never sever myself from the rush and heave of the external world—what have I lost?

I believe that I have lost something of my inner self. By inner self I mean that part of me that imagines, that dreams, that explores, that is constantly questioning who I am and what is important to me. My inner self is my true freedom. My inner self roots me to me, and to the ground beneath me. The sunlight and soil that nourish my inner self are solitude and personal reflection. When I listen to my inner self, I hear the breathing of my spirit. Those breaths are so tiny and delicate, I need stillness to hear them, I need aloneness to hear them. I need vast, silent spaces in my mind. Without the breathing and the voice of my inner self, I am a prisoner of the world around me. Worse than a prisoner, because I do not know what has been taken away from me, I do not know who I am.

The struggle to hear one's inner self in the noise of the Wired World might also be thought of in terms of private space versus public space. Public space—the space of people and clocks and commerce and deadlines and cellular phones and e-mail—is occupying more and more of our physical and psychic terrain. But the truly important spaces of one's being cannot be measured in terms of square miles or cubic centimeters. Private space is not a physical space. It is a space of the mind. It is "soul space," to use a phrase from Margaret Wertheim's book *The Pearly Gates of Cyberspace*. It is the domain of the inner self. When Dante makes his great journey through heaven and hell in *The Divine Comedy,* he moves not only through physical space but also through spiritual space. He visits immaterial realms of good and evil, beauty, truth. No wonder his companion and guide is the poet Virgil. Poets are masters of the inner self. In earlier centuries, physical space and soul space were united in a whole way of being in the world, of understanding the world. That dualism and wholeness are what I have lost.

Sometimes I picture America as a person and think that, like a person, our entire nation has an inner self. If so, does our nation recognize that it has an inner self, does it nourish that inner self, listen to its breathing in order to know who America is and what it believes in and where it is going? If citizens of that nation, like me, have lost something of our inner selves, then what of the nation as a whole? If our nation cannot listen to its inner self, how can it listen to others? If our nation cannot grant itself true inner freedom, then how can it allow freedom for others? How can it bring itself into a respectful understanding and harmonious coexistence with other nations and cultures, so that we might truly contribute to peace in the world?

IT IS A WARM spring day, and I stand in my classroom at the Massachusetts Institute of Technology, one of the world's great temples of technology. A freckle-faced student has just opened the large swinging window to allow some fresh air to waft into the room. I've had a schizophrenic career at MIT, teaching both physics and creative writing. Indeed, that split has followed my life's double passions in the sciences and the arts. Today, as usual, my students wander in late to class, eating bagels and pizza slices, wearing cutoff jeans and shorts and T-shirts and halter tops, complaining about some difficult problem on their physics or chemical engineering homework. My students are so bright, so quick, so eager to take their training into the world, and every one of them assumes, without question, that faster and more equals better. Hasn't that been the guiding assumption since the Industrial Revolution, that all developments in technology constitute progress? According to this view, if a new optical fiber can quadruple the transmission of data, then we should develop it. If a new plastic has twice the strength-to-weight ratio of the older variety, we should produce it. If a new automobile can accelerate at twice the rate of an old model, we should build it. MIT and many other institutions of science and technology do indeed have good departments in the arts and humanities, with the intention of graduating well-rounded human beings, and yet do not challenge the basic supposition: technology equals progress. "Progress" is some kind of ordained imperative of our species, an abstract conception of evolution, an inevitable direction of development like the increase in entropy, the future.

Centuries ago, technology was first and foremost associated with improving the quality of life and the human condition. (I use the word *technology* here retrospectively. In fact, the word did not enter the public vernacular until the founding of the Massachusetts Institute of Technology, in 1861. Early technologists and scientists might have called themselves craftsmen or engineers or natural philosophers.) One of my heroes in the history of science is Francis Bacon, whose *Novum Organum* (1620) proposed that nature could be understood only by careful, firsthand observation, as opposed to the acceptance of knowledge handed down by prior authorities. In his *The New Atlantis* (1627), Bacon envisioned a kingdom of science and technology, much of it unheard of at the time, that included living chambers where the air is treated for the preservation of health, the perfection of agriculture and the development of flowers and plants especially for medicinal use, glass lenses developed for "seeing objects afar off, as in the heavens and remote places," the study of sound and the creation of devices "which set to the ear do further hearing greatly." In this utopian kingdom, called Salomon's House, three "Benefactors" were charged with sifting through the experiments of all the house scientists "to draw out of them things of use and practice for man's life and knowledge."

Soon after Bacon, the development of technology became part of a major Western intellectual theme called "progress." Progress was centered around the notion that human beings were inevitably advancing to a higher plane—socially, politically, intellectually, scientifically, and morally. In France, Marie Jean Nicolais Caritat Condorcet's *Sketch of the Intellectual Progress of Mankind* (1795) proposed the concept of the "infinite perfectibility" of humankind. In England, the influential sociologist and philosopher Herbert Spencer attempted to synthesize the physical and social sciences and argued that a fundamental law of matter, "the persistence of force," inevitably brought about complexity, evolution, and progress in all things, cosmic and human alike.

In this grand idea of progress, which took on almost mythic proportions in the eighteenth century, intellectual progress was represented most notably by the theoretical discoveries of Isaac Newton and his sweeping laws of motion. The laws of gravity, discovered by the mind of man (Newton), governed everything from the orbit of the moon to the fall of an apple. Material progress was nowhere better symbolized than in James Watt's remarkable steam engine, the centerpiece of the Industrial Revolution. Power looms, for example, enabled textile workers to perform at ten or more times their previous rates and reasonably promised to raise the standard of living and relieve the exploitation of factory workers, as well as to increase the wealth of nations. Concern for the human condition was central in these developments. Technology in the service of humanity. On this score, I've always been inspired by the attitude of Benjamin Franklin—another of my heroes—inventor and scientist, statesman, philosopher, complete human being. Franklin refused to patent his many inventions for private profit because he felt that citizens should serve their society "freely and generously." In his famous *Autobiography* (1791), after giving a tedious account of his new invention for improving street cleaning, Franklin writes, "Human felicity is produced . . . by little advantages that occur every day." For Franklin and many other scientists and technologists of his day, the human being always came first.

LEO MARX, the distinguished historian of American literature and traditions and my colleague at MIT, occasionally joins me for lunch. Leo's landmark book *The Machine in the Garden* (1964) examined the way that the American self-identity, defined since early days by pastoral themes and images, has been confronted with and reshaped by the advent of technology. Leo is now in his mid-eighties. He still has most of his hair, and his striking blue eyes still look back at me with a penetrating clarity. As we sit on a bench with our cheese-and-turkey sandwiches, he gently suggests how I might think about technology and other large forces of the day. In his articles, Marx says that sometime in the mid-nineteenth century, the intention and direction of technology changed. Technology went from a means to humanitarian progress to an end in itself. The idea of progress, which had once meant an improvement in the human condition, became equated directly with technology. Progress was technology, technology was progress.

According to Marx and other historians of technology, at least two developments in the mid-nineteenth century helped change the nature and perception of technology. First, some areas of technology began to evolve from the individual-oriented "mechanic arts," like glassblowing and woodworking, to large, depersonalized systems, like the railroad. Secondly, these vast technological systems became hugely more profitable than any previous technology in the history of the world, offering great personal wealth to their creators. Technology became an instrument of the powerful enterprise called capitalism.

The earlier, mechanic arts were characterized by the skill of a small number of individuals and often a direct, personal contact between producer and consumer. By contrast, technological systems were large, amorphous organizations of machinery, people, and bureaucratic structures, with many levels between producer and consumer. Each railroad, the largest new technology of the nineteenth century, required thousands of workers, tracks laid for hundreds or thousands of miles, many stations, layers of bureaucracy and management, huge outlays of capital. (Compare this with the cell phone networks of today.) No longer was technology a humanistic activity, with its principal purpose to improve the quality of life. This turn of events led Henry David

Thoreau to make one of his more famous and witty remarks: "We do not ride on the railroad; it rides upon us."

An example that Marx uses to illustrate his point is a speech given by Daniel Webster, one of the foremost orators of his day, at a dedication of a new railroad in 1847: "It is an extraordinary era in which we live. . . . We have seen the ocean navigated and the solid land traversed by steam power, and intelligence communicated by electricity. Truly this is almost a miraculous era. . . . The progress of the age has almost outstripped human belief; the future is known only to Omniscience." Aside from the reference to Omniscience, the tone does not seem dissimilar to some of the early speeches and writings of Bill Gates or Larry Ellison or Gordon Moore. Nowhere in these words is there any reference to the quality of life, or human happiness, or the social betterment of humankind.

I look at my bright young students, so full of life, and wonder whether they can slow down enough to think about the purposes of their studies, think about what is truly important to them, as individuals and as members of a society.

MY INVESTIGATIONS turn to capitalism, possibly the most powerful organizing force in the world of today, certainly in the United States of America. I am not surprised to learn that capitalism has helped redirect the thrust of invention. And I even wonder: perhaps capitalism has always fueled the fires of technology, even Watt's steam engines and power looms. Capitalism lives on product, and no human creation has yielded product with such high efficiency as technology. More precisely, capitalism lives on profit, but the products of recent technologies have often translated into profits. Railroads, airplanes, telephones, automobiles, electric ranges and blenders, vacuum cleaners, dishwashers, microwaves and refrigerators, televisions and radios, Walkmans and CD players and video players, humidifiers, cell phones, copying machines, fax machines, personal computers—all have been gold mines for capitalism, returning great monetary gain to the inventors, creators, producers, and distributors.

As a consumer, I have benefited like most people from these rapid developments. I purchased one of the first Hewlett-Packard pocket calculators in the early 1970s and have owned a succession of powerful desktop and laptop computers ever since. I am a member of a two-cell-phone family. I have an electric garage door opener. I watch videos at home. I use the Internet to keep up with friends in other countries. I have certainly benefited from the advances in technology. But I have also paid a heavy price. And that price is what I most want to understand. That price, and even my personal benefit, are not of direct concern to capitalism. The first goal of capitalism is not to improve society and its members but to maximize the personal wealth of the capitalist. This goal is both the great strength and the great weakness of capitalism. I take out my copy of Adam Smith's *The Wealth of Nations* (1776), the bible of capitalism, and read: "It is not from the benevolence of the butcher, the brewer, or the baker, that we expect our dinner, but from their regard to their own interest. We address ourselves not to their humanity but to their self-love."

A good illustration of the relentless way in which capitalism and technology operate together is in the production-consumption work cycle of modern business. In the 1950s, academics forecast that as a result of new technology and increased productivity, by the year 2000 we could have a twenty-hour workweek. Such a development would be a beautiful example of technology at the service of the human being. In newly formed institutes of "Leisure Studies" and in such books as *Mass Leisure* (1958), experts pondered how Americans would spend their impending leisure time. More family vacations? More time for sports? More movies? More reading, more attendance at musical concerts or stage productions or art galleries?

According to the Bureau of Labor Statistics, the goods and services produced per hour of work in the

United States has indeed more than doubled since 1950. Half of the forecast was correct. However, instead of reducing the workweek, the increased efficiencies and productivities have gone into increasing the salaries of workers. Managers, desiring more and more profit, have found it against their interests to shorten the workweek or to stitch together part-time positions. Workers, for their part, have generally not lobbied for fewer hours but rather have used their increased efficiencies and resulting increased disposable income to purchase more material goods. As mentioned earlier, per-person consumption in the United States, in real dollars, has more than doubled since 1950. Indeed, in a cruel irony, the workweek in America has actually lengthened. The sociologist Juliet Schor, in her important book *The Overworked American* (1991), found that the average American worked 160 hours longer each year in 1990 than twenty years earlier. And that increase in working time cuts across all income levels. More work is required to pay for more consumption, fueled by more production, in an endless, vicious circle.

And what is it that we are consuming so voraciously, what impels us to work faster and longer hours, even in the face of higher efficiency? What are these burning material needs, when Americans have become wealthier and wealthier, more than doubling their real income per person in the last fifty years? One of the methods of capitalism is to create demand for its products, even when that demand does not previously exist. I was astonished to read this aim so unabashedly spelled out by Charles Kettering, a major inventor and executive of GM Research Labs. In 1929, at the beginnings of the automobile industry in the United States, Kettering wrote in *Nation's Business* magazine that business must create a "dissatisfied consumer" and "keep the consumer dissatisfied." A more recent example of the same idea was voiced by the economist John Kenneth Galbraith as he surveyed the future of capitalism in an ever-wealthier America. In his book *The Affluent Society* (1984), Galbraith writes that in modern America, production will have to "create the wants it seeks to satisfy." In short, a large part of our consumption is what we are told to consume, told that we need. And the cycle continues.

So it seems that we are running round and round like hamsters on the wheel of capitalism, production, demand, consumption, and work. Instead of slowing down the wheel, increased productivity has only sped it up. Instead of creating breathing spaces in the workweek, increased efficiency has caused us to work faster and longer. In this maze of counterintuitive results, it is hard to tell cause from effect, effect from cause. But the larger import seems clear. Ever since the physician George Beard commented in 1881 that "American nervousness" had increased since the invention of the telegraph, the pace of daily life has been set by the speed of communication and business. Everything in our lives has become faster, more hurried, more urgent. I cannot help but recall the first lines of William Wordsworth's poem, which is prescient in the way that artists often can divine the future:

> The world is too much with us; late and soon,
> Getting and spending, we lay waste our powers;
> Little we see in Nature that is ours;
> We have given our hearts away, a sordid boon!

MANY PEOPLE in the United States, both in intellectual forums and in daily conversation, have begun to express a fervent desire to slow down their lives, a sense of being trapped in a world that they cannot control. The word *helpless* is often repeated. For many of us, the practical difficulties of changing our work conditions and life rhythms are indeed enormous. Living in the Wired World as we do, are we then helpless to create private spaces and silences to contemplate our inner selves? Are we helpless to disconnect from the network?

I do not think so. In an odd way, my growing understanding of the vast forces that shape modern life has only increased my resolve to counter those forces, to build a parallel universe for my inner life and spirit. I am convinced that such an interior world is both possible and necessary. And, here, I disagree in part with two distinguished technology visionaries, the American co-founder of Sun Microsystems, Bill Joy, and the French philosopher and sociologist Jacques Ellul. Joy, in his provocative essay in *Wired Magazine,* "Why the Future Doesn't Need Us," argues that the world is being taken over by machines. Ultimately, Joy says, we humans will be the machines of the machines. Joy's prediction, which has much sympathetic resonance, is just too extreme. Although technology is proceeding at a dizzying pace, I believe that the human mind will always have control of itself. And since the human mind has a degree of infinity and imagination unlikely to be matched by a machine for a very, very long time, I don't think that we will become the machines of the machines.

Ellul, in his *The Technological Society* (1954), claims that technology and technical thinking have torn apart our world and rebuilt it into a rigid and unthinking society. Technology, according to Ellul, has so transformed our culture that "the human personality has been almost wholly disassociated and dissolved through mechanization." In Ellul's view, the technological mentality, the mentality of efficiency and production, is so pervasive that we have "no intellectual, moral, or spiritual reference point for judging and criticizing technology." I don't agree with Ellul for the simple reason that I did indeed have the moment of awareness that I described at the start of this essay. I did become conscious of the life I was living in the Wired World, I did remember the silences and inner solitude that I had experienced as a child, I did remember my places of stillness. I am writing these words.

A critical element, it seems to me, is awareness. In particular, becoming aware of the choices we have. Some of those choices are visible, some are not. Every day each of us decides, consciously or unconsciously, what to buy from the marketplace, what machines to have in our offices and homes, how to use those machines, when and how to communicate with the outer world, how to spend our time, what to think about. When do we unplug the telephone? When do we take our cell phones with us and when do we leave them behind? When do we read? When do we buy a new microwave or television or automobile? When do we use the Internet? When do we go out for a quiet walk to think? These decisions may seem petty and trivial. But at stake in these hundreds of daily decisions is the survival of our inner selves. We have choices, but we must become aware of these choices.

I do not believe that needed changes can be mandated from the top down. First, the underlying malaise of the Wired World is not primarily economic or legal. Rather, it is philosophical, psychological, and spiritual. And second, invididuals have different priorities, different values. It is the slowness and silence and privacy for reflection on those values that we must regain. While it would be helpful for governments to enforce new laws—such as that cell phones are forbidden in restaurants or that all businesses must provide six weeks of vacation for their workers (as in some European countries) or that all public and corporate spaces are required to have noise-free zones or that transactions in the stock market must have a twenty-four-hour delay—none of these mandates can by themselves alter attitudes of self. Changes in philosophy of life come about slowly, and at the level of the individual.

A comparison is the institution of slavery. Slavery has existed in all parts of the world since the earliest recorded history. The involuntary servitude of one class of individuals to another is sanctioned in Mosaic law and described in the Old and New Testaments. Despite their life of high culture and refinement, the ancient Greeks not only permitted slavery but also organized much of their society around it. For millennia, slavery was accepted as part of the natural order of things. Without question, it was believed that some human beings were naturally inferior to others, could rightfully be owned by others. Even many of the founding fathers of America, such as Thomas Jefferson, counted slaves among their possessions. The various antislavery laws in

Pennsylvania in the late seventeenth century, in Denmark and France in the late eighteenth century, and in England in the early nineteenth century did not stop slavery. What ended slavery was the gradual recognition by individual members of human society that slavery was dehumanizing, not only to the slaves but also to their masters.

In creating the Wired World and the mentality that goes with it, we have unintentionally imprisoned ourselves. That imprisonment has happened slowly and unconsciously. Our manacles are subtle and invisible as air, but they are real. Although the regaining of our freedom and the reclaiming of our inner selves will take time, it is possible.

FINALLY, I RETURN to the present, the moment of my questioning. The Wired World, for good and for ill, is the world that we live in. Capitalism and technology, for good and for ill, are here to stay. But, as potent and pervasive as these forces are, I do not think we can blame them for the absence of privacy and silence and inner reflection in our lives. We must blame ourselves. For not letting my mind wander and roam, I must blame myself. For allowing myself to be plugged in to the frenzied world around me twenty-four hours a day, I must blame myself. Only I can determine my personal set of priorities and values, reflect on who I am and where I am going, become aware of those many small decisions I make throughout the day. The responsibility is mine. Understanding that the responsibility is mine is a kind of freedom in itself.

Thoreau framed the problem well a century and a half ago when he said that we must produce better dwellings "without making them more costly; and the cost of a thing is the amount of what I will call life which is required to be exchanged for it, immediately or in the long run." Somehow, each of us must figure out how to measure the "life," our personal life, our inner self, that we exchange for each piece of technology or scheduled project or public connection. This accounting may have to be done item by item, hour by hour, but I believe that it must be done and it can be done only by the individual. Only individuals can measure their own values and needs, their own spirit, their own story of life.

(2002)

ACKNOWLEDGMENTS

In this age of high specialization, it is not easy to pursue interests that cut across traditional disciplinary lines, and I am extremely grateful to the institutions that have allowed me to do so, especially the Massachusetts Institute of Technology. For the same reason, I am also grateful to the editors who have encouraged me and commissioned particular writings. I thank Alison Abbott at *Nature,* Robert Silvers at *The New York Review of Books,* Dennis Overbye at *The New York Times,* Jim Miller at *Daedalus,* and Margaret Hancock at Hart House of the University of Toronto. I also thank Priscilla McMillan for sharing with me a number of unpublished and critical documents for my essay on Edward Teller.

I am grateful to my friend and editor of long standing, Dan Frank, and to my friend and literary agent of long standing, Jane Gelfman. Finally and always, my loving family, Jean, Elyse, and Kara.

Some of the essays in this collection have previously appeared in the following: "Metaphor in Science": *The American Scholar,* vol. 58, number 1 (Winter 1988–89) • "Inventions of the Mind": *Brick,* number 50 (Fall 1994) • "A Sense of the Mysterious": *Daedalus* (Fall 2003) • "Words": *Nature* (October 18, 2001); adapted from "The Physicist as Novelist" in *The Future of Spacetime,* by Timothy Ferris, Stephen Hawking, Alan Lightman, Igor Novikov, and Kip Thorne (W. W. Norton, 2002) • "The One and Only": *The New York Review of Books* (December 17, 1992) • "The Contradictory Genius": *The New York Review of Books* (April 10, 1997) • "Megaton Man": *The New York Review of Books* (May 23, 2002) • "Portrait of the Writer as a Young Scientist": *The New York Times* (May 9, 2000) • "A Scientist Dying Young": *The New York Times Magazine* (March 25, 1984) • "Prisoner of the Wired World": *Living with the Genie,* edited by Chris Deser, Alan Lightman, and Daniel Sarewitz (Island Press, 2003); originally presented as the 2002 Hart House Lecture, University of Toronto.

ABOUT THE AUTHOR

Alan Lightman was born in Memphis, Tennessee, and educated at Princeton and the California Institute of Technology, where he received a Ph.D. in theoretical physics. An active research scientist in astronomy and physics for two decades, he has taught both subjects on the faculties of Harvard and MIT. Lightman's novels include *Einstein's Dreams*, which was an international best seller; *Good Benito; The Diagnosis*, which was a finalist for the National Book Award; and *Reunion*. His essays have appeared in *The New York Review of Books, The New York Times, Nature, The Atlantic Monthly, The New Yorker*, and other publications. He is currently adjunct professor of humanities at MIT.

- BLUE BERRIES
- CHICKEN
- SWISS CHARD
- RADIANT RED